Southern Gold

by

Deryle Wesley Chavis

VANTAGE PRESS
New York Washington Atlanta Hollywood

FIRST EDITION

Published by Vantage Press, Inc.
516 West 34th Street, New York, New York 10001

Manufactured in the United States of America
Standard Book Number 533-02503-6

Acknowledgments

In obtaining material for this treatise, gratitude is conveyed to the Georgia Department of Archives and History, Maryland Department of Geological Survey, Arkansas Geological and Conservation Commission, Alabama Department of Conservation and Natural Resources, Georgia Department of Natural Resources, Tennessee Division of Geology, South Carolina Division of Geology, Mineral Resources Division of North Carolina, Virginia Division of Mineral Resources, New Jersey Bureau of Geology and Topography, Vermont Department of Libraries, Dahlonega Chamber of Commerce, the information service of the U. S. Mint, Connecticut Department of Geology and Natural History, and the Massachusetts Department of Public Works.

Indebtedness should also be expressed to the South Carolina Public Library System, especially at Florence, S.C., and to Professor J. R. Butler for their fine cooperation in preparing this edition.

Contents

Preface

The prosperity of modern man depends upon the ready supply of minerals in quantity, together with the energy and intelligence he puts upon the utilization of the total of all his natural resources. Heretofore, progressive changes in the daily life of the average person were slow; indeed, an average family in the early 1800s lived very much like those of hundreds of years before. Rapid technology that promotes fast social changes could not have been possible without sufficient funding made available through gold resources, the initial bulk of which, at least until 1848, were supplied by Southeastern mines and miners. Any nation, including the U.S., with little or no gold supply will progress slowly, if at all.

Beyond mineral and financial requirements for technological advancements, increased currency for everyone, with a minimum of devaluation, i.e., multiplication of purchasing power, is an essential factor in raising living standards. In that respect, individual and cooperate gold mining efforts in America's Southeast in the early days of our nation's history produced a sizable inventory of gold reserves that supplied money nationally, and internationally as well.

Nevertheless, the complexities of such ventures nowdays are such that an average person, before deciding on any gold investments of any kind, should look far beyond mere interest and romanticism and gain a rational understanding of money and mining efforts so that he may not be lured into a decision that may adversely affect him and those around him.

Southern Gold

A Brief History

Everyone is familiar with the glamorous years of early western American mining and explorations, yet few people are aware that mining techniques were developed in the East long before they appeared elsewhere in the U.S.; moreover the gold reserves and base metal resources of early America were derived from the mining efforts of pioneering Easterners.

In North America, artifacts indicate that the precolonial Indians had little or no use for gold, the hunger for the precious metal being introduced with European explorations. There are unofficial reports of Western mining by Indians and adventurers in pre-Revolutionary War days. It is also believed that three North Carolina mines (the Oliver in Gaston County, the Dunn in Mecklenburg County, and the Parker in Stanly County) were worked prior to the War for Independence. Additionally, Thomas Jefferson, in the earliest printed reference to gold in Virginia (1782), reported on the recovery of 0.4 troy oz. of precious metal from a four pound gold bearing rock found on the north side of the Rappahannock River about four miles below the falls.

The first officially recorded instance of a gold discovery was in 1799 in Cabarrus County, near Concord, North Carolina. When, Conrad Reed (Reid), a boy of twelve years of age and the son of a German immigrant, found a seventeen-pound nugget while walking through Little Meadow Creek on the Reed Plantation. For some time the nugget sat on the floor and was used as a doorstop because no one believed that a piece of natural gold could be that size. Two years later, a Fayetteville jeweler confirmed it as being the precious metal. From that discovery, with panners and miners appearing everywhere, the year 1801 is credited as the beginning of gold mining in America, with North Carolina as the first state.

Little Meadow Creek as it appears today, about where the first nugget was found

According to *Historical Sketches of North Carolina 1584–1851* by John Wheeler, chunks of gold ranging from twenty-eight pounds to one pound with a total of 115 pounds of large pieces were found in Little Meadow Creek from 1803 to 1835. No estimate was made as to the gross overall recovery of the average large size of sixteen pounds down to the finest of pieces. An eyewitness is reported to have said that "the whole of the creek for nearly a mile was very rich in gold." The Reed Mine, now a state historical site, yielded mostly placer gold, although underground workings at one time numbered fifteen separate shafts. The value of the rocks from the veins was nearly $1.00 per pound near the surface, but dropped sharply as depth increased. The record now stands at a total of 153 large nuggets of various sizes, the largest of which, twenty-eight pounds, valued at $800 in 1803, would at today's gold price fetch well over $50,000 for the same weight.

Tunnel at Little Meadow Creek

With discoveries becoming commonplace, more and more people turned to prospecting and mining, resulting in the discovery of the precious metal throughout the Southeast. Prospecting activity and thoughts of sudden wealth were probably intense, for in 1828, with the discovery of a rich deposit in Georgia, the nation's first gold rush had begun. The place, Dahlonega, called "the Gold Capital of Georgia," was named after the Cherokee word *tau-lon-eka* which means "yellow money." Reportedly, pieces of gold were so abundant on and near the top of the ground, particularly after a shower of rain, that small boys would frequently pick up pieces of the metal weighing up to one pennyweight (1.6 grams).

As the word spread and the "fever" grew, new settlers began arriving daily. Soon all roads leading to the town were lined with huts or cluttered with shanties, enlarging the area to ten to fifteen thousand miners within a few miles, followed, of course, by gamblers, con men, and other gaudy accessories of an expanding gold strike. An early traveler is supposed to have said that "they came afoot, on horseback, in wagons, acting more like crazy men than anything else."

By July of 1829, adding to an already fiery turmoil, the stampede had "filled the diggings with a wild and lawless population." Because of the mountain fastness, the Cherokee country had long been a refuge for law-breakers. Subsequently, Georgia passed a law requiring settlers on Indian lands to take an oath to uphold the constitution and the laws of the state, hoping that such measures would at least restrain disobedience. Hostilities were commonplace, now fanned by gold seekers and settlers; unrestrained atrocities were often committed on and bordering the Indian lands, for the federal government, in sanctioning Cherokee sovereignty over their territory, placed such activities .beyond the jurisdiction of the Georgia authorities.

One group of outlaws taking advantage of the frontier disorder, a rascally bunch known as the "Pony Club," preyed upon early settlers along the Old Federal Road, an improved trail from western Tennessee to Augusta, Georgia. At that time, it was considered by many to be "the darkest and most dangerous place in the Cherokee Nation."

Rugged and boisterous frontiermen were well described by Professor Girard Troost, Tennessee's first state geologist, in 1831, suggesting the origin of the name "Hot Water" for a district of the well-known Coker Creek placers of southeastern Tennessee, saying, "In a small strip of land enclosed nearly all around by ridges and spurs, is the Hot Water washing, which has produced much gold; and it was monopolized by a parcel of men who sustained their right by fist and club, perhaps also by dagger, bowie knife and rifle, so that those who attempted to wash for gold at that place got into hot water, and when returning showed more bloody noses, bruised limbs and black eyes, than gold."

Besides villains and gold hunters, complications from the wilderness itself, the Indians, and the missionaries who assisted and guided the Cherokees, so swelled that federal troops were required to restore order. They remained a short while and left only after an assurance from the state government that order could be maintained by the Georgia Guard. Trespassing and land grabbings continued, law defiance remained, residency permits were refused, and, in general, the problem persisted.

The odds were against the Cherokees, for, as civilized and advanced as they were—having developed an alphabet (Sequoya, George Guess, 1825) and printed a bilingual newspaper in English and their native language *(Cherokee Phoenix);* their overall organization with a written constitution (1827), recognized by the federal government as a separate nation; together with the persistent aid from the missionaries, who for nearly a century had counseled the Indians; and a growing nationwide sentiment for them—at least three circumstances remained adamant, namely, opposition from two government officials and the lure of gold.

Georgia's new governor, Wilson Lumpkin, and the president of the United States, Andrew Jackson, were calloused and unsympathetic toward the needs of all the Indians. Even a ruling by the U.S. Supreme Court in favor of the Cherokees could not deter the single aim of removing the whole Indian Nation from the East entirely. When speaking of gold, and especially about the Cherokees, the state of Georgia's and sympathizer President Jackson's viewpoint, though not shared by all of Congress and certainly not by the Supreme Court and other interested persons, was that "the state was being wrongfully deprived of property of untold wealth."

In the 1831 U.S. Supreme Court decision in *The Cherokee Nation vs. the State of Georgia,* it was declared that the

Indians were neither a foreign nation nor citizens and had no right to bring suit, but were wards of the federal government and that their lands were a result of occupancy rather than ownership. Privately, Chief Justice John Marshall supported their position, but could not act since the suit was not properly before the Court.

Later, missionary Samuel A Worcester, not being an Indian, but acting in their behalf, properly presented his case before the U.S. Supreme Court. It was decided that the acts of the state of Georgia were void, meaning that they should not meddle in Indian affairs. Allied with the president, Governor Lumpkin ignored the findings and things continued as they had been. President Jackson, in learning of the verdict, is reported to have said, "Marshall made his decision; now let him enforce it."

The federal "colonizing" program enacted in the 1817 and 1819 treaties granting 650 acres of land and citizenship to any Cherokee who could administer his property was appealing to many, for, soon, they had worked the land and advanced so much that representatives of the government, in making trips through the Cherokee country, were astonished at such rapid progress.

Most likely, with sufficient time, the Indians would have been much as anyone else; unfortunately, though, time was not to be allowed for the pressure from gold seekers and the demands for money strained the situation in all respects. Not being allowed to testify against a white man and being prohibited from gold digging while the invading whites appeared to have a free hand fostered severe antagonism among the Indians, and often reprisals, yet the state of Georgia, seemingly oblivious to the Indians' predicament, went ahead with its program of frontier settlement.

Thus, by 1831, the state ordered Cherokee lands to be surveyed so that by the following year it could be laid out into ten new counties. Following that, the mineral lands were divided into forty-acre lots for public domain settlement, to be determined by lottery. Even though prospectors had already encroached onto the territory and settled along the Etowah and Yahoola Rivers, whoever partook in the drawings could, by chance, draw either rich or poor ground. Apart from the gold producing area, however, regular lots (non-mineral lands) composed of one hundred-forty acres each were sold similarly with every Cherokee head of a family allowed a reservation of that amount, but no Indian was given a land title.

By 1834, even without a proper treaty, whites were allowed to occupy Indian holdings and, as awarding of lands increased, trading in real estate became as profitable as gold mining itself; prices increased as the gold deposits continued to produce even greater quantities of high grade ore. Clearly, the Indians were being maneuvered into a compromising situation with the U.S government forced to participate. With animosity from the state capital and pressure from Washington, it was clear that some sort of settlement had to be agreed upon.

A small faction, claiming tribal consent, was persuaded to sign a treaty (1835) in which the Cherokees would receive five million dollars and a title to land beyond the Mississippi River. Legal transaction or not, President Jackson accepted it as such, delegating the moving order to General Winfield Scott, who harshly rounded them up and by 1838 had begun the western migration. It's certain that, of some 14,000 Cherokees who started out from Georgia, the Carolinas, Tennessee, Virginia, and Alabama, some 4,000 died on route. The slow westward trek of some 700 miles in nearly five months was so marked with misery, disease, and death that it has appropriately been called the "Trail of Tears." A small band escaped, eluded the law and the army, and finally, after several years as fugatives, gained a legal footing on a reservation in western North Carolina, where they have afterward kept alive their traditional manners and customs. Part of their story is depicted in the annual pageant "Unto These Hills" played each summer near the town of Cherokee, at the Cherokee Reservation in North Carolina's Great Smokey Mountains.

Initially, Georgia's northern frontier settlements followed the gold producing areas. At one site, Nathanial Nuckalls settled, built a tavern, and established the beginning of the town of Nuckallsville, later Nuckellsville, and finally Auraria, a rival of the nearby town of Dahlonega. From that area at the time of the Colorado gold rush (1858), the Russell Party, famous for early Colorado gold discoveries, joined in the Georgia exodus to Cherry Creek as part of the "Pikes Peak or Bust" stampede. Applying what they had learned of real estate values in gold mining regions in Georgia to the new Western lands, they settled on the southwest bank of Cherry Creek, naming their village Auraria in honor of the mining town in Georgia of the same name, creating a great deal of business competition with their neighboring rival, St. Charles, on the northeast bank of the same river, then the western reach of the Kansas Territory. It was renamed Denver City in honor of the territorial governor, James Denver, and consolidated with Auraria in 1860. The next year Colorado became a federal territory with Denver designated as its permanent capital in 1867. The Colorado rush, less than a generation after the devastating manpower depletions of the California stampede, and just before the Civil War, left the south bereft of laborers and craftsmen.

The Dahlonega surge is generally considercd the first genuine gold rush in America, yet it was only a highlight in an already mounting tide of fermenting activity and booms throughout the Southeast. A promise of wealth made some more eager; for others, a fair wage at the mines meant a steady livelihood; businesses flourished, with some stores accepting unminted gold in payment for supplies simply by "weighing out the price." Use of foreign money dwindled, but currency shortages still persisted. As an expedient measure, private companies were authorized to mint coins, resulting in fifteen private mints in addition to the U.S. Mint at San Francisco, having operated in California by 1854.

In relation to the number of early Americans and the economy of the time, a substantial amount of gold was

Chronology of Gold Rushes

Concord, N.C. (Little Meadow Creek)	1801	Pikes Peak, Colorado (Cherry Creek)	1858
Dahlonega, Georgia (Duke's Creek)	1828	Klondike, Alaska-Can. (Rabbit Creek)	1898
Sutter's Fort, Cal. (American River)	1848		

produced by North Carolina and later by several other Southern states, yet few people are aware that one of the most serious problems confronting the commercial interests of early America was the failure of the government to provide a sufficient volume of circulating coins. The fault was not in the supply of metal to the mint or of the mintage, but rather in interfering circumstances disrupting the expected flow of new coinage into the channels of trade.

It turned out that Spanish dollars (pieces of eight, eight bits) and their fractional parts continued to circulate with official sanction until 1857, at which time it was declared that foreign coins could no longer be used as legal tender. Worn Spanish coins of reduced weight and value were easily exchanged for U.S. silver, meaning a loss almost as soon as they were minted of most of the new dollars originally designed for free circulation. Gold coinage witnessed a similar fate in that it began about even in value to world gold currency, but by 1799 European trade centers had increased the gold content of their coins, meaning the undervalued American coins tended to flow out of the country, mainly through international trade exchange, resulting in gold coins' being rarely seen in general circulation for some time after 1800. The total money supply apparently was just as critical after 1801, when North Carolina began mining and shipping to Philadelphia for conversion, as before, indicating improper mintage and currency controls.

A common practice of unscrupulous individuals was known as "clipping," a term describing the shaving of small slivers from a coin's surface. A modified version of illegal metal recovery was to place several money pieces in a sack and shake and tumble them, reproducing natural wearing, and saving the precious metal dust caught in the bag. These deceptions of weight changes were easily done because the Spanish and other currency metals were irregular shaped, called "cobs," having been struck from crude silver bars with rough dies. The true value could be determined only by weighing the coins, regardless of their face value. Merchants and others dealing with quantities of money during the colonial period and for a long time thereafter, when working with questionable coinage, reserved a pair of small scales for just such a purpose. Indeed, coins shipped to England in exchange for European goods were acceptable only by weight. Few people realize that clear prints and milled edges, together with a near-perfect roundness, were designed to prevent profiteers from recovering metal at the expense of the currency.

For these reasons, and more, gold from the Southern states faded into manufacturing and trade channels, resulting in a negligible change in spending money between 1794 and 1834. It has been estimated that there was less than one American coinage piece for each person in the country in 1830. This era of nondescript currency included America coinage, several varieties of standard and underweight foreign coins, and notes from "wildcat banks" flooding the country before 1830, becoming more common than silver coins. Such banking practices lacked a firm foundation, and often resulted in financial ruin for the investors and the bank as well.

History, in repeating itself, has shown time and again that overinvestment, inflated dollars, and expanded credit has led to economic chaos. Continental notes, called "continentals," were backed by the newly formed Congress of the United States, faithfully pledging to redeem the paper bills in Spanish dollars (silver coins), even fining citizens for not accepting the paper currency. With excess printing, confidence faded, rapidly depreciating the value and causing higher and higher prices for everything purchased. Early economists were learning that paper money is only as good as the precious metal that is redeemable to the bearer on demand.

By 1779 the paper dollar value had dropped to three cents. Later, with the value even less, some economists saying 50,000

Reed Mine historical marker

Gold Hill historical marker

Crystallized quartz rock from the Carolina Slate Belt, Cason Oldfield, South Carolina

Panned gold from the King's Mountain Belt, Smyrna, South Carolina

to 1, a common phrase for something worthless became, "not worth a continental." State currency, French aid, and the reversion to metallic coins, a tangible item universally negotiable, after some time restored confidence in the medium of exchange. The financial shambles of 1837 is similar, occurring at a time of seemingly unending prosperity, in that it was mainly caused by overexpansion of business, unredeemed extension of loans, and overspeculation in real estate. Clearly, America's bouts with paper money and inflation made gold and silver the only substances in which the people had total confidence.

In 1934, things were different. After several unsuccessful efforts to stop the Depression, drastic measures, were adopted such as abolishing gold coinage and prohibiting hoarding; inflating dollars through devaluation (gold changed from $20.67 per Troy ounce to $35 per Troy ounce); followed by an embargo on gold exports and the formal abandonment of the gold standard; and inaugurating an attempt to raise commodity prices through controlled inflation, i.e., managing the currency and stimulating the economy by switching exports from gold to goods. These led to improved conditions, meaning that a nation can prosper without gold currency, but only through sufficient regulations.

Of the seven U.S. branch mints that have existed at various times since the Coinage Act of 1792 that authorized a mint at Philadelphia for gold and silver, three have been in the Southern states. Additionally, two Germans, Christopher (Christian) and Augustus Bechtler, prompted by influential citizens to ease transaction boondoggles, operated a private mint in North Carolina that equalled or excelled federal standards in fineness and weight; they also had the distinction of striking at that mint America's first gold dollar. In that same vein, Templeton Reid, a goldsmith and assayer, produced limited coinage in Georgia before moving to California. Some of the Bechtler dies are in the State Hall of History at Raleigh, N.C., and the old press on which the coins were struck is kept at the Museum of the American Numismatic Society in New York.

Later, the desire of Eastern business interests for gold coinage, plus a wish to develop newly discovered gold mines in North Carolina and Georgia, were probably the chief forces leading to the passage of the Coinage Act of June 28, 1834. It provided that a branch of the U.S. Mint would be established at the city of New Orleans, for coinage of gold and silver, at Charlotte, N.C., for gold only, and at Dahlonega, Georgia, also for gold only. Southerners were to have their own mints to coin Southern gold; still, at least since 1803, they had been sending gold and silver to the Philadelphia mint for conversion into circulating money. Moreover, the entire gold supply for federal coinage from 1804 to 1827 was derived solely from North Carolina mines. South Carolina began supplying in 1827, and Georgia in 1828, with Alabama, Tennessee, Virginia, and Maryland all at various times adding to the nation's total gold supply.

U.S. Mints

Gainsville, Ga.		
Templeton Reid		\$ 2.50
Private Mint		\$ 5.00
1830- ?	Gold Unknown	\$10.00
Rutherfordton, N.C.		
Bechtler, C. & A.		\$ 1.00
Private Mint	Gold	\$ 2.50
1830-1852	\$3,625,840.00	\$ 5.00

Southern Mints and Mintages

Charlotte, N.C.				\$ 1.00
	Gold			\$ 2.50
1838–1861	\$5,059,188.00			\$ 5.00
Dahlonega, Ga.				\$ 1.00
	Gold			\$ 2.50
1838-1861	\$6,106,569.00			\$ 3.00
				\$ 5.00
New Orleans, La.				
	Gold	Silver		
1838-1861	\$40,318,615.00	\$ 29,890,037.00	3c	gold dollar
			half-dime	\$ 2.50
1879-1909	\$ 8,385,557.50	\$220,066,498.10	10c	\$ 3.00
			25c	\$ 5.00
			50c	\$10.00
			silver dollar	\$20.00

Federal Coinage, Southern Mints

Charlotte, N.C.
(C)

Gold	*Number*	*Face Value*
$5.00	881,027	$4,405,135.00
$2.50	217,966	544,915.00
$1.00	109,138	109,138.00
	1,208,131	$5,059,188.00

Dahlonega, Ga.
(D)

Gold	*Number*	*Face Value*
$5.00	1,107,211	$5,536,055.00
$3.00	1,120	3,360.00
$2.50	197,850	494,625.00
$1.00	72,529	72,529.00
	1,378,710	$6,106,569.00

New Orleans, La.
(O)

Gold	*Number*	*Face Value*
$20.00	818,775	$16,375,500.00
$10.00	2,361,089	23,610,890.00
$ 5.00	923,725	4,618,625.00
$ 3.00	24,000	72,000.00
$ 2.50	1,209,263	3,023,157.50
$ 1.00	1,004,000	1,004,000.00
	6,340,852	$48,704,172.50

Silver		
$ 1.00	187,111,529	$187,111,529.00
50c	80,234,674	40,117,338.00
25c	60,343,000	15,085,750.00
10c	68,079,906	6,807,990.60
half-dime	16,246,550	812,327.50
3c	720,000	21,600.00
	412,735,661	$249,956,353.10

Mint Museum of Art, housing mint relics at Charlotte, N.C.

Another provision of the 1834 act was to reduce the weight of standard gold (monetary parity), in effect placing Americans on a gold standard, greatly benefiting trade and finance. Before the mints opened, though, the law of 1837 revised and standardized the mint and coinage laws, assuring the nation of a substantial coin supply. By 1853 such measures, with adequate mintage, for the first time in our nation's history, had produced an ample supply of fractional coins for general circulation.

With the beginning of the War between the States, the Southern federal mints closed; thereupon, the facilities continued functioning by striking Confederate currency. The spontaneous and impromptu Civil War showed the economists to be ill-prepared, for the mined gold and silver stockpiles for the most part had already been converted and distributed, leaving them with a mininum reserve. Additionally, nearly all of the gold and silver mines closed, shifting to base-metal production. Thus coin mintage soon stopped for lack of metal, and Confederate paper money suffered dearly from insufficient backing. Later, as hostilities ended, with the exception of the Dahlonega facility, the Southern branches of the U.S. Mint become assay offices and precious metal depots. Only the New Orleans member resumed coinage, thereby minting until 1909, and afterwards serving as an assay office and depot until 1942.

Since limited restrictions had always been applied to money handling, at least fifty million tokens in more than 8,000 varieties were issued by a total of twenty-two states during the Civil War years. Following the National Banking Act of 1863, designed to restrain uncontrolled free banking, bank failures, and unredeemed notes, Congress abolished private coinage with another law on June 8, 1864, ending yet another glamorous chapter in American history.

Two major events grossly affecting Southern gold recovery were the Civil War and World War II, the latter relating to mines nationwide while the former was of concern mostly in the Southeast. By 1860, the manpower shortage caused by the flight to Western gold fields were one consideration; still, the Southern states, being largely rural and for the most part unaccustomed to mining, primarily engaging in agriculture and related industries, were estimated to be contributing about two-thirds of the goods exported to Europe. Historians sometimes express the idea that if the South at that time were a separate nation, it could have competed in every way with any other nation in the world, saying further that Southern gold played a role in behind-the-scenes negotiations to win England over to their cause.

What people were not so much aware of then, and later historians are learning, is that the North made financial maneuvers that, in effect, put them in control of the "purse-strings." It turns out that much of the export profit from the sale of staples went into the pockets of Northern businessmen through the cotton factorage system, indirect trade with Europe via Northern ports, and purchases by Southerners of

Northern manufactured goods. Other considerations include the fact that Northerners had invested heavily in Southern gold mining, up to five million dollars in the God Hill district of North Carolina alone. It turned out that more than sixty mining companies operated in North Carolina at one time or another between 1817 and 1860, and had attracted a total investment of perhaps 100 million dollars from Northern states and foreign countries.

Unfortunately, because of the low volume of individual mines, precious metal deposits created transient financial booms, notable exceptions being the Haile Mine in South Carolina and the Homestake Mine in South Dakota; consequently, the spurious activity quickly dwindles. By 1848 the precious metal mining industry in North Carolina at times employing as many as 30,000 men, and ranking second only to agriculture, waned so quickly that by 1860 it could hardly be called a major industry. Other more stable businesses, within twelve years, and before the Civil War, had markedly surpassed gold recovery in North Carolina (by far the largest precious metal producer in the Southeast) in yearly value, number of employees, and future productivity.

Major Industries in North Carolina in 1860

Rank	*Industry*	*Business*
1	Turpentine	1,526 establishments
2	Flour and Meal	639 mills
3	Tobacco	97 small factories
4	Lumber	330 sawmills
5	Cotton Textiles	39 factories

As the war effort mounted, gold recovery throughout the Southeast virtually stopped, agriculture and manufactured goods became limited, and accrued capital was quickly used up. During that time, unrestrained printing of Confederate and federal paper money was to create America's second runaway inflation (continentals were first). Greenbacks of the North depreciated to about one-third of their value. In

California, where both paper and gold were used interchangeably, the paper dollar declined to three to one, with prices tripling in terms of paper money, yet remaining the same in terms of gold. Clearly, the federal government, and certainly the Confederate states (with limited metal backing), had issued too much money in paper form.

No one knows the actual confederate mintage of the remaining stockpiles coined with federal dies, later struck with confederate designs; still, according to estimates in 1861, of the nation's total money supply represented by Western and Southern mining 275-300 million dollars was in gold, subsidiary silver, and minor coins, the South holding only 20 million in gold coins and about 18 million in subsidiary silver coins, the remainder being in the North and West. Bank notes, representing gold and silver deposits were in use, but no paper money, as U.S. currency had ever existed before 1861.

Because of disproportionate currency and of the extended war effort (activists thought it would end quickly and thus made short term planning) coins in the South soon disappeared into banks and elsewhere. With few taxes and credit based upon future agricultural productivity, the Confederate paper money that soon appeared began to have less and less value, having decreased to 2 cents worth of gold for each dollar by March, 1865. With virtually no real money in the South, the diminishing possibility of ever producing any, and the devastation that ensued during the war (several mines, including the Haile prospect were ravaged by federal troops), rebuilding processes were necessarily slow continuing well into the next century.

Reconstruction and renewed interest in gold recovery could not revive the spirit of the "gold fever" days. Sporadic efforts dependent upon scientific engineering characterized the mining era from the Civil War to World War II. Except for occasional good fortune from minimal prospecting for fun and profit, labor costs and production factors since the Second World War have all but ended Southern gold mining. Besides scant individual endeavors, all that remains of the "stampede" and mining fury that once engulfed north Georgia is a gold resort that partakes in old gold mining activities engineered in the tradition of the early days and directed toward recreation for tourists through old gold mining practices with many relics of those historic days displayed at the Dahlonega Gold Museum, housed in the old village courthouse.

The Appalachian goldfield is believed to have been formed from paleozoic sediments when tectonic forces from the east or southeast folded and raised the Appalachian Mountain Range, overturning many close folds toward the west. Gold ore therein is usually thought of as occurring in crystalline rocks within folliated igneous and metamorphic formations on the eastern slope or foothills (piedmont) of the elevated hardened alluvium. Along the trailing edge of the Appalachians from southern Maryland to northeastern Alabama, the gold-bearing formations normally range 10 to 100 miles wide by 700 miles long, and are broken up into a number of near-parallel belts, individual veins of which may or may not carry any precious metal ore.

Included in the system are seven states; Virginia, the two Carolinas, Georgia, and the contiguous portions of Maryland, Tennessee, and Alabama. Arkansas is the only Southern state with an isolated precious metal deposit, albeit of such low assay and high dispersion that successful recovery operations from those ores would be dependent upon large-scale production and adequate metallurgical facilities.

Tennessee, while yielding scant recovery from its placers and veins in the southeast portion of the state, salvages 100 to 200 ounces yearly as a by-product from its complex base metal ores of the celebrated Ducktown mining district. Louisiana is often considered the farthest reach of the Southern states, while others sometimes question the inclusion of Texas as part of the sectional group. If included, most of its gold production would be from veins in Llano County, with some placers, especially near the coast, and as a by-product of its copper and silver ores. Those recoveries, together with some working from central Texas and the Edwards Plateau, have been estimated to be probably less than one million dollars.

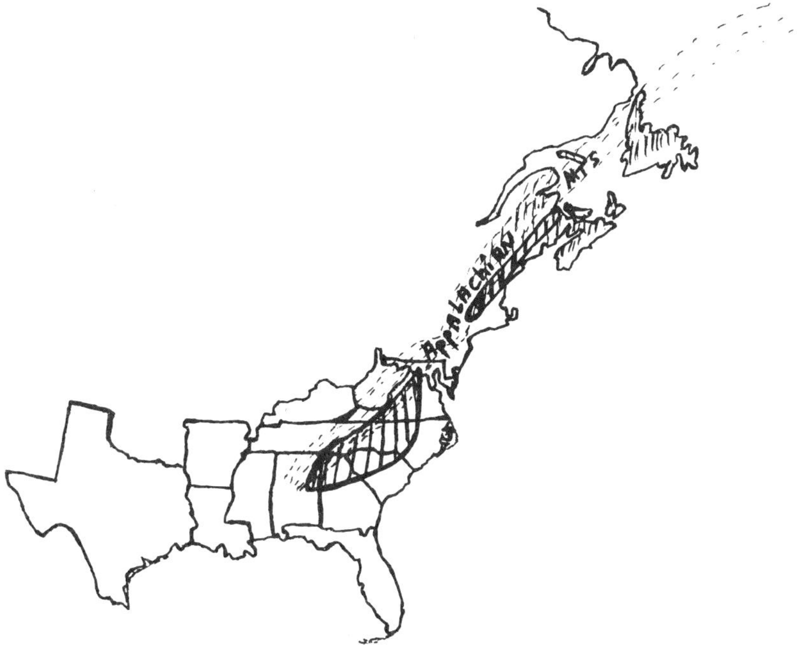

Approximate Range of the Appalachian Gold field

An overall view, including the small amount of gold known to be present in Nova Scotia and Newfoundland together with the same pattern of continuing rock formations, may, in theory, indicate that gold-bearing minerals from fissure veins or igneous magmas may occur in the submerged portion of the Eastern seaboard. If that is the case, it may, in part, explain the well-known high concentration of oceanic gold resources off the coast of Newfoundland. Placer and veinal metal recovery from New England and the upper piedmont of Pennsylvania, although a by-product, reinforces the supposition of geographically extended continental gold resources. Interestingly, iron (magnetite, pyrite, chalcopyrite) ore from the Cornwall Mine in Lebanon County, Pa., in 1943 produced, from copper concentrates, seventy-seven percent of the total Eastern gold output.

Northern Recovery Through World War II
(Troy oz.)

Pennsylvania		*Vermont*
1942	2,499	—
1943	2,218	—
1944	2,115	100
1945	1,518	104

Conversely, the narrowing of the gold belts at the two extremes and the occurrence of numerous quartz veins in upper Maryland and in New Jersey, in some cases carrying pyrite, but with no gold occurring in any of them, or any indication of a potential occurrence, indicates that precious-metal-forming conditions may terminate, or at the very least deepen, so as to be concealed in that region. Some individuals, however, maintain that a minute trace of gold does occur in one of New Jersey's conglomerates, but that this does not indicate local gold-bearing minerals.

The rock formations that sometimes weather and disintegrate to form the familiar gravel placers and other secondary deposits are nearly always quartz minerals, a special form of the well-known white or discolored flint (chert) rock. It is usually crystallized, denser, and may carry several other valuable minerals even if no gold is present.

Massive quartz with localized crystallization, Carolina Slate Belt

Conditions favoring the formation of quartz veins carrying gold or other minerals are geochemically interpreted as involving cracks and faults deep underground and near an intruded magma (undeveloped volcano) bearing an ample supply of heavy elements. It may be that the abundance of barren veins in gold-producing regions, even though they have all the necessary qualities, being formed the same time and in the same way as those considered to be rich ores, could have formed extensive mining districts if the original voids had been sufficiently close to ample metal-bearing magmas to allow hypothermal percolations to extract the minerals and alloys, to sort and concentrate them, and to contain them in a silica body as it filled and hardened in the gap between the severed formations. Indeed, Tennessee's only granitic formations occur along the same path as the narrower, shorter (about fifty miles long) gold-bearing alluvium and veinal deposits, the common country rock of that region being metamorphosed shales and sandstone. After the veins formed, sufficient time would be required for erosion and removal of overburden to expose the ores so that they may be found and worked.

The dispersed gold deposits of Arkansas occur in the east-west, intensely folded and faulted Ouachita Mountains in the central and western part of the state, extending into Oklahoma. Quartz crystals and veins are a very common feature of the Ouachita region, yet they bear little gold and silver. Lack of igneous intrusions at the time of the vein formations may account for the deficiency since less than one-tenth of one percent of the total area of the state contains orogenic formations, whereas granite and other magmatic formations frequently occur in many places all along the piedmont of the East Coast. Besides veins and placers, gold also occurs in massive formations such as in the Haile Mine (called replacement deposits), and disseminated in a porphyritic manner, not chemically mixed or alloyed, but rather as fine particles dispersed in native and pyritized metals such as copper zinc, and lead.

Apparently as the Applachians rose, relieving the compressing strain, a larger portion of the squeezing pressure developed on the eastern side, allowing cracks and faults to develop near the end of the deformation and at the close of the paleozoic era. An accompaniment of extensive, intrusive magmas allowed percolating minerals to deposit gold-bearing ores along dikes and seams (fissure veins) closely paralleling the main axis of the mountain range.

The resulting precious-metal districts (belts) are likely to be scattered and some distance apart. Georgia has some 500 mines and prospects within four such separate belts, plus a few irregular deposits, the most famous district, of course, being the Dahlonega vein cluster, containing intermittent deposits over an area 150 miles long and one to five miles wide.

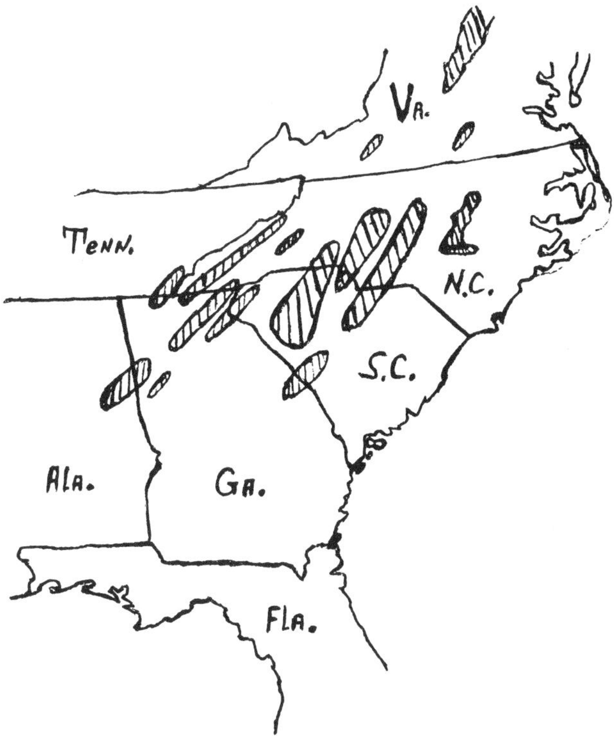

Approximate Range of Gold Belts in North Carolina, South Carolina, Georgia, Alabama, Tennessee and Portions of Virginia

Individual belts are usually made up of a great number of veins or ore bodies running parallel to each other and conforming in dip and strike to the gneiss and schist, the country rock, varying from a fraction of an inch to several feet, and often continuing without interruption for long distances. Findley Ridge, Georgia's most productive single district, south and east of Dahlonega and rising a few hundred feet above the surrounding countryside, once contained irregularly distributed quartz stringers and veins four to six inches wide by one and one-half to two inches thick of well-formed shoots extremely rich in coarse gold.

If the miners of that region had continued as they apparently intended to do, much more metal and possibly several types of other minerals could have reached a high degree of development; but, unfortunately, the California and, later, the Colorado, rushes depleted the area of skilled artisans. The 10,000 or so residents who had flocked to Auraria to serve the mining community had nearly all gone and left it a virtual ghost town by 1860.

Noticing the exodus and results thereof, Matthew Stephenson, assayer of the Dahlonega Mint, in 1849 spoke to a crowd gathered at the courthouse, in an unsuccessful effort to stem the tide of western migration from Georgia's goldfields. Pointing to Findley Ridge, he said, "There's millions in it." Mark Twain, upon hearing of the speech, wrote the now-famous version in his book *Gilded Age*, "There's gold in them thar hills."

Most of the better-quality areas have been at least moderately developed, reaching their greatest output by 1845. Present-day explorations in the Southeast reveal some high-grade ore of low quantity and large quantities of low-grade ore, the possibility of discovering rich veins of substantial volume being unlikely, though not impossible. For the most part, future gold-mining developments depend to a great extent upon low-grade engineering and metallurgical applications to show a profit.

The underlying rocks of the piedmont are highly deformed and complexly faulted, thus the minerals exposed in the East

Quartz vein, Carolina Slate belt

Coast goldfields indicate that they were formed under a variety of conditions of heat and pressure, and are distinctly different from the deep, barren roots of the mountain system. The distribution and exploitable potential may not be so evident from customary methods of surface exploration, since deep weathering may have destroyed the minerals to some depth, disguising their presence and preventing disclosure of their potential wealth. Clearly, in every sense, it is fallacious thinking to say that Southeastern gold resources no longer exist and should be regarded as a forgotten industry.

The three leading districts of the Appalachian fields are Findley Ridge at Dahlonega, Ga., the Haile Mine at Kershaw, S.C., and the mines at Gold Hill, N.C., and, except for the single Haile deposit, all were separately operated mining projects clustered about a small area. In large measure, a 100-mile radius each of Charlotte, N.C., and Dahlonega, Ga., would encompass the most productive precious-metal areas of the Southeast.

The Three Most Productive Districts of the Appalachian Gold-fields

Site	*Tenor*	*Recovery at $35/oz.*
Haile Mine, Kershaw, S.C. (1829–1942)	0.15 oz./ton	$6,500,000
Gold Hill, Gold Hill, N.C. (1842–1935)	1.5 oz./ton	$5,600,000
Findley Ridge, Dahlonega, Ga. (1828–1935)	Unknown	Believed to be about the same as Gold Hill

Various recovery districts of the more than 660 locations formed within six gold belts that have been mined at one time or another, spread over nearly 1,000 square miles, have made North Carolina the East's most gold-productive state; yet, it was not until 1842, with washings at the head of a drain, that mining operations commenced at the now-famous Gold Hill.

NE—SW Gold-producing Belts of Central and Western North Carolina

Eastern Carolina Belt
Carolina Slate Belt
Carlotte Belt
Kings Mountain Belt
South Mountain Belt
Western Belt

Hard-rock veins were discovered and worked so extensively that by 1843 the camp town was named Gold Hill in honor of the mines and miners. Situated directly above some of the mine tunnels and serving the gold district of about one square mile, in 1848 it had five stores, one tavern, four doctors, and 1,000 laborers in the nearby mines, some of whom were Cornish workers from Cornwall, England. Their experience in English coal mine shafts made them the best skilled help of the time; and, by applying their knowledge and techniques, greatly assisted in early mining developments.

The large, well-developed, vertical veins yielded an average assay of $30 per ton at $20.67 per Troy ounce, which, even by today's standards, would qualify as one of the richest deposits. Water pumped from the mine was used for steam power to drive rock crushers and mechanical separaters for a twenty-four hour day. Very little amalgamation was used, for the ore was of such quality that minimal metallurgical work was required. Limited antebellum facilities confined operations to a moderate depth, while the postwar period, with specially developed equipment, allowed a vertical depth of 820 feet, with drifts and tunnels branching off in all directions at different depths. Similarly, Georgia's most productive single operation, the Creighton (Franklin) Mine, reached 950 feet on the incline of a vein, and except for the war was worked continuously from 1840 to 1909, when it was untimely closed as a result of flooding by the Etowah River.

Collectively, the Gold Hill operations form a group that has been more extensively developed than any other in the Carolinas. Except for the Civil War years, the area was continuously worked until 1915, some tailing and surface work continuing until 1935. Besides gold, $700,000 in copper was also recovered from the same district. Apparently, the resulting mineralization is part of a shear zone developed along a fault that separates schist on the east from intrusive granite rocks on the west. Another district along the same belt contains several silver veins (Silver Hill) that were worked almost as extensively as the veins at Gold Hill.

As a single mine, the most extraordinary of the North Carolina operations was the Old Howie (Condor) prospect. The largest in Union County, N.C., it operated until 1942 with an average assay of 0.833 Troy ounces of gold per ton. In the beginning and near the surface, it yielded as much as ten ounces per ton, and two and one-half ounces per ton even at extended depths. Until granite was found deep underground, it was thought to have been formed peculiarly different from all the other well-formed deposits.

Lesser yields from Maryland, Alabama, and Virginia probably reflect the diminished width and tapering off of the far extremes of the gold belts. The first discovery in Alabama was made near Jemison in Chilton County, followed by placer mining at Arbacootchee, Cleburn County, at the headwaters of Dime Creek. That deposit and several other alluviums were worked until the Western rushes; unfortunately, though, only sporadic and mostly unsuccessful attempts at placer and veinal mining have produced minor amounts of recoveries since then from its more than 100 prospects and mines.

By far, Virginia's James River district, probably because of its size, surpasses two other gold areas in that state, a few irregular deposits and the Virgilina district of south-central Virginia. As with most Southern states, the 1840s represented its greatest productive era. The south-central mining areas have the added value of copper minerals while the gold-pyrite belt of the James River district appears to yield more gold. Extending northward, the belt crosses into Maryland, forming its Great Falls gold-quartz district.

Gold Production in Virginia

early 1800s–1845	39,898 Troy oz.
1846–1860	35,094 Troy oz.
1861–1875	4,250 Troy oz.
1876–1890	5,530 Troy oz.
1891–1905	3,918 Troy oz.
1906–1920	1,824 Troy ox.
1921–1935	1,474 Troy oz.
1936–1950	6,627 Troy oz.

Applying the $35-per-Troy-ounce value to 2,597,963 Troy ounces, the official Southeastern recovery up to and including the World War II years, as reported in the *Minerals Yearbook*, reveals a total recovery value of $90,928,705. The current price of gold, of course, would make it about five times that figure. The number of ounces reported, though, represents an official minimum, for the metal was often marketed to non-government mints and assay offices, commercial refineries, and private commerce. In that regard, it is estimated that at least fifty-five to sixty million dollars worth was recovered from North Carolina mines before 1860, while from 1828 to 1848 thirty-five million dollars worth is believed to have been obtained from the Dahlonega district alone. Additionally, Clyde Pittman, in his book *Death of a Gold Mine*, suggests that occasional nuggets and metal pockets at the Haile operation may not have been included in mining reports. It is also suspected that Gold Hill yielded far more metal then the official record implies.

Official Southern States Gold Production (1799–1945)
Source: *Minerals Yearbook* (1943–45)

State	*Duration*	*Fine oz.*	*Value at $35/oz.*
Alabama	1830–1945	49,494	$1,732,290
Georgia	1830–1945	870,526	30,468,410
Maryland	? –1945	6,102	213,570
North Carolina	1799–1945	1,164,588	40,760,580
South Carolina	1829–1945	318,825	11,158,875
Tennessee	1831–1945	20,870	730,450
Virginia	1828–1945	167,558	5,864,530
		2,597,963	$90,928,705

Inasmuch as South Carolina follows North Carolina and Georgia in total gold production, its Haile Mine, about three and one-half miles northeast of Kershaw, S.C., in Lancaster County, is the largest single producer of gold, not only of the Carolinas, but of any mine in any state east of the Mississippi.

That particular deposit began being developed when Col. Benjamin Hale (Haile) leased twenty plots of land for placer work. Processing of ore began in 1837, followed by underground developments of shafts and tunnels by 1880. Processer, methods, and output were so well developed that management and recovery from the mine gained worldwide recognition through the years 1832–1917.

Haile Mine, as the Northeast Section exists today

Because Pyritic minerals contained finely dispersed precious metals, processing problems increased, production slowed and almost stopped until, in 1888, Dr. Carl Thies, a German mining engineer and scientist in the field of metallurgy, fresh from the Phoenix Mine in North Carolina where he developed the "Thies Chlorination Process," applied an improved operation for low sulfide minerals.

Erosion of production and intermitent operations followed the 1908 boiler explosion until 1934, when a new company, Haile Gold Mines, Inc., began its cyanide treatment of 100 tons of ore per day. By 1937 they had attained a capacity of 400 tons per day that continued to mid-1942, that year producing fifty percent of the Southern recoveries, with sixty-nine percent the year before. Since then there has been no reported gold production at that site. For the whole state, only intermitent precious metal activity, leasing, and small-scale explorations have occurred since then, but no significant recoveries. Interestingly, second in Southeast gold production at the beginning of World War II was the fabulous Condor Mine (Old Howie), near Waxhaw, Union County, N.C., in which it contributed fourteen percent in 1941 and twenty-seven percent in 1942.

It turns out that, of South Carolina's more than 130 mines and propects from its two recognized gold belts, recovery in the state has been dominated, especially since 1880, by the Haile operation, recovering an estimated $6,500,000, and that the mine's activity alone reflects the fluctuations in the total amount of gold produced in the state. It is located in the central northeastern district, called the Carolina Slate Belt. Those crystalline rocks, no more than twenty miles wide, extend into North Carolina for some distance being disguised in places by overlapping coastal plain sediments.

Among other prospects, the Haile and Brewer Mines are located at the lower extreme of the belt, about eight miles apart. Besides native bismuth, fine-grained massive topaz occurs at the Brewer Prospect, the only such deposit in the East, having been rumored to have yielded some crystals to early explorers. Moreover, that deposit is atop one of the highest hills, among massive boulders, yet very nearly two-thirds of its gold was recovered from placer deposits.

South Carolina's Three Largest Gold Producers
(valued by $35/oz.)

Haile Mine, Lancaster County	$6,500,000
Dorn Mine, McCormic County	900,000
Brewer Mine, Chesterfield County	450,000

The importance of Eastern gold was considerably lessened by the Western discoveries. But, remember, when precious metals were found in other places, no one had to apply trial-and-error methods to develop mines and metallurgical processes; the explorers already knew the most modern mining techniques as well as tried and proven ways of winning the metal from its ores. This really means that the Southeast ventures, in addition to giving an emerging nation a solid foundation of gold reserves, producing all of the American coinage metal and a moderate portion of the gold bullion, before 1848, also developed the mining and mineral expertise to complement Western mineral exploitations.

With gold from California, the U.S. became the world's leading gold producer. By 1905, South Africa had become the leader and has held that position ever since, presently followed by the U.S.S.R., Canada, the U.S., and Australia, in that order. The impressive recovery era of the 1840s was rivaled 100 years later by the early 1940s where gold production reached an all-time high. The year 1941 witnessed the largest output of any single year at the Haile Mine—$542,780 at $35 per ounce. The small total production from all the America gold mines in 1942 reflects the closing of most of them that year by federal order.

The Gold Mining Limitation Order L-208, issued by the War Production Board on October 8, 1942, declared that all "nonessential" gold mines would close and cease all gold mining activities. "Essential mines," those lode mines handling less than 100 tons per month and placers washing less than 100 cubic yards also in one month in 1941, were exempt from the ruling, provided the operator of the small-mining interprise held a serial number issued under Preference Rating Order P-56.

Regulations and loss of artisans caused by the nationwide manpower measure to direct gold mining employees to base-metal production, coupled with rising costs and diminishing supplies, devastated the Eastern precious-metal mines; for, since their closing, only minimal exploration and recovery has been reestablished, with North Carolina being the only Southern state that routinely recovers some gold from genuine gold ores. By-product precious metals from Tennessee continue to be supplied, and apparently will be for some time to come.

Certainly some deposits still exist for the forced shut-downs were not because of lack of ore; indeed, the mines were producing at an all-time high. After consideration, it was probably decided that cost factors versus potential recovery were too great; for, to reopen a mine after some length of time, whatever the reason for the closing, requires restaffing of specialized personnel and refurbishing of equipment, processing operations, and particularly the mines themselves, often requiring complete overhaul of their internal structure. Besides historic value, the Appalachian gold region then, with its more than 1500 mines and prospects, should not be excluded from the nation's future gold-producing capabilities.

Weights and Values

Of all the uses of gold, it is estimated that more than ninety percent of the entire world's supply is used for monetary purposes by governments and central banks, most of it being kept as a reserve in the form of bullion to stabilize paper currency. That metal, being the only commodity universally accepted in payment of debts, though not legal tender anywhere and used to back paper money in most of the countries of the world, is therefore the standard of exchange in international trade.

In a redeemable sense, America is not on the gold or silver standard, but rather a "gold bullion" or modified "gold standard" system. The gold dollar, still the standard unit of value, is the parity at which the secretary of the Treasury is required to maintain all forms of money issued or coined by the U.S. government. Until 1934, the standard dollar contained 25.8 grains of gold at 900 fine (90%), resulting in 23.22 actual grains, or exactly one dollar's worth of gold at $20.67 per Troy ounce.

The 1934 to 1972 era describes the gold content at 15 5/21 grains at ninety percent or 13.71 actual grains and exactly 1/35 of a fine ounce of gold, meaning thirty-five dollars would purchase one Troy ounce, establishing the new price of gold at $35 per Troy ounce. In 1973, thirty-eight dollars would be required to buy the same amount of the precious metal; and, with more changes developing as time goes on, the carefully designed system of money management and controlled inflation is rapidly becoming unrestrained.

Paper money, then, is not directly backed by any metal, but by the power of the Federal Reserve System, the U.S. Treasury maintaining gold reserves for international exchange and for payment of foreign debts when paper currency is unacceptable to those with whom we trade. In effect, today's dollar represents an unredeemable government debt that any debtor can legally require his creditor to take in payment of a debt, but still is useful and valid only because of its acceptance by the general public. Monetary gold, then, is being used, but remains beyond the reach of any unauthorized citizen.

The Free World's reserve stock of gold is now in much less proportion to that of ten years ago, the balance of payments and other factors for a long time being in favor of foreign countries. Outflowing gold bullion, of course, very rarely leaves the country, being transferred on accounts and moved from one storage compartment in the gold vault to another.

Monetary Gold Reserves
(millions of dollars)

	U.S.	*Free World*	*Percent*
1962	16,057	41,470	38.7
1963	15,596	42,310	36.8
1964	15,471	43,060	35.2
1965	13,806	43,300	31.9
1966	13,235	43,205	30.6
1967	12,065	41,600	31.4
1968	10,892	40,905	26.6
1969	11,859	41,010	28.9
1970	11,072	41,275	26.8
1971	10,206	44,742	22.8
1972	10,487	45,000	23.3

Public demand with political and financial response over the American gold losses became so great (1968) that emergency measures were taken to protect the U.S. monetary gold position. However, recent diminishing mineral supplies and the energy crisis may again plunge the U.S. Treasury into deep financial trouble, meaning more borrowing and extended credit from the world bank.

Values of Gold Imported and Exported
(thousands of dollars)

	Imports	*Exports*
1962	150,932	380,962
1963	44,414	203,784
1964	40,888	422,744
1965	101,669	1,285,097
1966	42,004	457,333
1967	32,547	1,005,199
1968	226,263	839,159
1969	236,906	12,287
1970	237,464	37,790
1971	283,947	51,249
1972	357,689	63,053

The Federal Reserve System, supervised by a board of governors (seven men appointed by the president with the advice and consent of Congress, one of whom is appointed chairman) who closely coordinate and control twelve privately owned regional reserve banks that in turn regulate twenty-four branch and member banks, reports to Congress on the lending rates and money supply these banks lend to the familiar commercial banks. Their controls also apply to checking accounts, since it represents "artificial" money, contributing to about four-fifths of the U.S. currency supply.

Formerly, the available supply of money was based on the amount of gold in the Treasury; nowadays, a formula is being used that is based upon the ideas of printing volume economists feel are necessary to maintain normal business activities. Objectives, motives, and methods are to be well studied, for changes in the consumers' supply of credit and money influences the aggregate demands for goods and services, affecting the nation's output, price level, and employment, all schools of economics agreeing that increasing the money supply and bank credit causes price inflation and creates an inflationary boom. The system is also charged with replacing lost and unusable coins and bills, printing each year more than one billion bills. Greenbacks change hands so fast and so often that the average life of a paper note is only about one year (less for dollar bills).

It is convincingly argued that the "prosperity" of the Western nations is not genuine, but rather built upon borrowed money. The desires of the American public have grown faster than their ability to pay, resorting to ever-increasing debts, in some cases to the extent that they have expanded the money supply so much that the claims far outweigh their tangible physical assets. Clearly, they borrow on the expectation of perpetual plenty, when in reality expenses may be growing faster than income.

To secure a hedge against expanded credit and inflated dollars, several people have turned to hoarding gold, others to investing in silver or diamonds. While speculation is for gamblers and financiers, some average Americans, on a small scale only, are retaining some native and processed gold, others indulging in part-time prospecting and mining. Recovery of native metals is often financially superior and nearly always more rewarding in a psychological and physiological sense than merely purchasing processed elements. Regardless of trickery, purchased gold, items made of gold, or any other substance, always contain metals of considerably less value than the cost of the item at the time of purchase.

American currency does not respond to the fluctuating gold sales of the international market as in the case of countries that are on such a standard. Of the few places that precious metals are bought and sold, the prices vary, yet American money remains very nearly the same value. It turns out that one such emporium, the London Gold Market, transacting a large portion of the world's sales, has become one of the most important international precious metal exchanges.

England has trafficked in gold for centuries, but it was not until 1919 that the current market structure was established. The five firms of which it is composed surround themselves in a mystique unparalleled by any other market in the world. Members of the market, sometimes called "bullion brokers," deal with their clients and each other as principals rather than as brokers.

Although much interest concentrates on the price-fixing, marketing takes place throughout the day. Currently, with no formal rules, the twice-daily gold price-fixing is transacted by representatives from the five member firms in an office of one of the members. Each agent maintains telephone contact with his own organization and bargains with the others until the number of buyers and sellers are about evenly matched at a certain price. That gold value is then "fixed" and is posted as such. Their transactions, though undisclosed, are believed to range into several hundred million dollars yearly for each firm represented.

Historically, gold has been the most sought after metal, at least since the first legendary mining venture of Jason and the Argonauts who sailed the ship *Argo* in search of the Golden Fleece, the sheepskin in which the inhabitants of Colchis supposedly caught gold dust too fine to be recovered otherwise, making the precious metal recognized as valuable since antiquity. That famous legend was based on an actual expedition (about 1200 B.C.) to seize gold from a region later called Armenia, where sheepskins aided in gold recovery from river sands.

Substances not particularly esteemed by man have value totally dependent upon supply and demand at a particular time. Gold, on the other hand, is an object which men during all ages and in all parts of the world have valued more highly than any other metal; and, because of its unique qualities, has served both as a dependable medium of exchange and as a storehouse of value since the early dawn of history.

Its universal acceptance as the most desirable element probably came about as a result of its durability and scarcity, and equally so for its color and functionality. Only one other naturally occurring metal, copper, has a distinctive color. Gold is durable almost to the point of being indestructable; also, it is inert in air, water, and in any single acid except selenic acid, which only tarnishes gold (a surface reaction). Aqua regia, a mixture of nitric and hydrochloric acid is the only effective acid solvent, functioning through the generation of nascent chlorine, forming gold chloride—a stable, water-soluble, extremely hygroscopic (deliquescent), orange powder.

Mercuric amalgamation, cyanides, and activated halogens are the only practical dissolvers of gold, all having been used at various times in mining operations to win the metal from its ores. The workability of the element surpasses any other metal, in that it can be alloyed to improve its hardness, melted, and shaped, yet retain its original beauty. Beaten or rolled into thin sheets, called gold leaf, one ounce will make 250 square feet, 300,000 of which are required to make a pile one inch thick. For decorative purposes, usually about 23K, translucent gold sheets may be as thin as 0.00001 millimeter.

Native gold, as it occurs naturally, is always alloyed with other metals, mostly silver with minor quantities of other elements. However, at certain places in the western U.S., various tellurides occur as the only natural gold mineral that is chemically combined. Occasionally some Appalachian veins or placers may assay gold at better than ninety-five percent pure, others characteristically being in the low nineties; scarcely any deposits, other than placer and stream gold from the upper eastern Carolina Slate Belt, which approximates about seventy percent pure, run as low as California lode gold, typically about eighty-five percent. It turns out that the Western metal, while less pure, may be more widespread and possibly with a larger total volume within a given area than its Eastern counterpart. Precious metals from the East Coast, on the other hand, while more likely to be purer, form in pockets even though the veins carrying it may be continuous for several yards or a mile or more. Such deposits, with some exceptions, prevent large-scale, highly developed operations, yet allow for individual discoveries that, by chance, may prove profitable.

Traditionally, among jewelers, the *karat* has been the quality of measure of gold purity, while fineness, also a purity measure, is used mostly by industry and coinage mints. Conversely, carat weight is a quantity measure of gemstones only, having nothing to do with precious metals or their purity. Even more confusing is the use of the apothecaries' weight (Troy system) as opposed to avoirdupois weights (standard system). An easily remembered approximation is that the Troy is always very nearly ten percent heavier than the standard.

Weight Relationship
(Quantity)

Grams	*Apothecaries (Troy Ounce)*	*Standard (Avoirdupois Ounce)*	*Carats (Gemstones only)*
0.200 (200 mg.)	0.008	0.009	1
1	0.032	0.035	5
28.4	0.91	1	142
31.1	1	1.1	155.5
373.2	12 (Troy pound)	13.2	1,866
453.6	14.6	16 (standard pound)	2,268

With some exceptions, the weight of a precious metal is given in Troy ounces, Troy pounds seldom if ever being used; additionally, the karat or fineness always accompanies a transaction, normally having the purity stamped into the processed metal object. Interestingly, a standard practice of the U.S. Mint is not to accept gold less than 200 fine (twenty percent pure).

Purity Relationship
(Quality)

Karats (metals only)	*Fineness*	*Purity*
24	1,000	100 %
18	750	75 %
12	500	50 %
6	250	25 %
1	41.7	4.2%

Most people are aware that, in the United States, there has never been any restriction upon individuals acquiring, retaining, or selling native gold. Processed gold, on the other hand, has only recently become unrestrained; still, everyone should know that trafficking in precious metals belongs to an exclusive association of speculators and financiers.

Local disposal of a small amount of any precious metal is a fairly straight forward transaction, but selling of a moderate to large quantity involves some particulars peculiar to valuable metals. The several well-advertized refining firms usually accept the metal as is, purify it, pay whatever they say the market value is at the time they finish the work for the weight they say they get, and charge their standard refining fee for a certain quantity of metal processed. Clearly, a novice should know what he is sending and the expected value of the recovery, for the processing cost at one refiner for silver is $15 per 100 ounces of impure metal, for gold twenty percent of the recovery value, and accordingly expensive for other precious metals. The refining costs, of course, are directly related to the difficulty of impurity removal, silver being simple and easy while platinum is complex and requires much more time and effort. More work and extra recovery steps mean more loss of metal; therefore, refiners, especially when working with low efficiencies, are the only ones who know exactly how much is recovered. Besides the cost, refiners almost exclusively work with metals only; compounds, in most cases, are unacceptable. For large sales, particularly of rare or unusual substances, some proof of ownership may be required.

A rich strike or large volumes are required to produce enough precious metal to exceed the refining cost sufficiently even though several placers contain a fair amount and virtually every stream in the Appalachian goldfields contains some gold. Unless an unimaginably rich deposit is found, a hand panner will never recover very much precious metal, for, even at a better panning site, he may recover only about three cents worth of gold each time.

Consider that a seasoned panner pans at the rate of two to three minutes per pan with each sixteen-inch pan holding about ten pounds; then, after six hours of continuous work, he would recover about four dollars for working nearly three-quarters of a ton at three cents per pan. These values are exaggerated, for current alluvial deposits very rarely exceed three to five dollars for a full ton any place, including California or Alaska. In that regard early prospectors and miners have been rated at one ton and about one dollar per day.

A gravel deposit, more than likely, is not altogether gold bearing, but has auriferous sediments only in a small part of it. A pay streak that transverses the gravels (old stream channel) that is about 300 feet long by 75 feet wide with an eight-inch average thickness hardpan (dry, packed gravels) would contain 558 cubic yards or approximately 890 tons of gold-producing alluvium (one cubic yard of gravel usually weighs 3,000–3,400 pounds). At five dollars per ton, no more than $4,000 dollars, for the metal only, can be expected from this deposit.

One ton of alluvium, representing one day's work for one panner

Regardless of the equipment, the precious metal yield efficiency will depend upon the metal's particle size. Very fine gold (gold dust) that will pass through a forty-mesh screen may contain as many as 4,000 colors per ounce of gold with a value of about one-tenth of a cent per color at $35 per ounce. Particles that small are claimed by some equipment manufacturers to be recovered at the rate of eighty to ninety percent. Flour gold is even smaller, averaging 1500 colors, and up to eight million individual pieces in some places of the Snake River in Idaho are required to make one cent. The flour type, being bouyant, sometimes floating, is usually what is lost in recoveries; however, as a partial compensation, when proper methods are applied, it is usually of relatively high purity.

While a hand panner's capacity is about one ton, with pick, shovel, and wheelbarrow, an individual, working alone, is able to dig up and scoop into washing and separating contrivances about three to five cubic yards per day, or, under optimum conditions, about eight tons, as the limit for a manual operation for one person. The wages obtained from such an operation in the early days was very nearly $1.00 per day, in many cases, less. Workable alluvial ore, then as now, may run as low as 0.01 ounces per ton. Present-day gold prices, about eight times the early value, meaning a low-volume miner would make $8 a day or less, does not allow an income from mining to be commensurate with today's living costs. Plainly, a better grade of ore must be found, or else, if that deposit is to show a profit, some machinery for larger volumes must be employed.

Disproportionate incomes from gold recoveries—such as the gold digger of 1830 in Tennessee who made $2,000 per day for three weeks; or the documented Parker Mine in North Carolina beginning as a placer, and later as veins yielding a rich alluvial deposit in three or four acres estimated then at $200,000 (about 10,000 ounces) mostly in the form of nuggets, the largest of which weighed eight pounds four ounces—do not describe the average individual's earnings.

Representative wages are depicted in the report (Geology of Tennessee) in 1869, in which Stafford, speaking of individual placer mines, says, "at first, in the richest localities industrious men could average $2.00 per day which was then, when wages were low, considered an excellent business. In a short time the average was reduced to $1.50, then to $1.00 and lower. About 50¢ a day to the hand can be depended upon with the additional prospect of greater yields at intervals."

By contrast, in the earlier days of antebellum gold mining, moderate to exceptional profits are reported to have resulted from the sporadic use of slave labor by miners and local farmers, with the result that, in Virginia at least, servitude formed the larger portion of the original work force. As an added note, some free blacks chose mining as a livelihood, some even owning and operating their own recovery facilities.

In 1830, one of South Carolina's most productive eras, the Brewer Mine, at times employing 100–200 miners, is said to have yielded $1.50 to $3.00 per day per person. Miners in those days had an agreement of profit sharing, but the gross profit was dependent upon how much each individual person was able to recover from the ground, himself. Assuming four tons each, $3.00 per day would mean that 0.036 ounces per ton at $20.67 per Troy ounce were recovered per day, per person. The realizable potential recovery, however, may be nearly twice that reclaimed by the placer miners at the Brewer mine, for some early observers have reported extreme losses through gross inefficiencies. This evaluation is very close, for Nitze and Wilkens, in an 1897 report, stated that the average assay at those diggings was about $3.00 a ton for the gravels and hardrock pyrite ores combined.

As salaried employees, the eighty cents a day Georgian miners were paid in the year 1900 was decidedly better than the $10 a month with room and board for laborers in North Carolina in 1860. A skilled worker at the Haile Mine in South Carolina in 1902 could make as much as $1.25 a day. The salary of Professor Troost, Tennessee's first state geologist, was said to be $500 per year, or about $1.90 per working day, meaning a person who was placer digging at $2.00 a day was making as much as a college professor or most any other citizen of that time. Currently, things are proportioned differently, inasmuch as the amount of precious metal to be recovered in some places now may be of the same assay as in the mid-1800s, yet the value of a day's yield in terms of modern purchasing requirements and in comparison to revenue from other occupations is such that any ideas a person may have of sudden wealth from gold mining will have to take on a totally different meaning from those of yesteryear.

The Prospectors

Overzealous profiteers bent on financial rewards from mineral developments sometimes forget that there are three distinctly different groups involved: prospectors, miners, and landowners. Usually they are each a separate entity; two may be combined, but rarely are they all three in one. Few people need to be reminded that the proprietors, since they have invested nothing, are more likely than anyone else to gain from any mining venture.

Miners are borderline and can lean far in either direction, depending upon the business aptitute of the management involved. Prospectors, on the other hand, being the impractical dreamers they are, especially if they have little capital resources and less mining ability, are destined to a life under conditions that only the promise of gold can make them endure. Moreover, they have always been a restless lot, ready to stampede to whatever mining locality is rumored to be rich, frantically searching and always hoping that their bonanza is just within arm's reach.

Although rather simple in design, a gold hunter's equipment—consisting of boots, magnifying lens, hammer, magnet, pick, shovel, two pans, and a steel mortar and pestle—will suffice in evaluating outcropping mineral deposits, rock veins, or auriferous gravels. Of the two, the gravelly placer is more easily found, easier to work, and, through its development, may lead to the vein deposits that originally gave rise to the colors.

In the strictest sense, many Appalachian gold sediments are not true placers, but are residual deposits (saprolites) that have resulted from in-place vein weathering followed by minimal erosion. In that regard, the richest portion would be near the surface and would become progessively leaner with increasing depth, forming many of the so-called surface mines. Several deposits of this type occur in Georgia and have been worked extensively from time to time.

Natural sorting and concentrating of precious metals is done more effectively and efficiently by a moderate stream, for swift water erodes badly and deposits little, while a sluggish stream, in depositing all the metals, tends to blend everything together in a type of loose conglomerate. Consequently, stream and dry land gravels (ancient riverbeds) can give erroneous results unless careful sampling and cautious interpretations are applied. In all cases, judicious prospecting, slow and patient work as it necessarily is, with a systematic procedure and knowledge of the formations, is certain to result in the discovery of any gold that may be present.

Descriptive terms such as "wealth seekers," "prospectors," "gold hunters," "argonauts," and "forty-niners"

Wolf Creek, in the Kings Mountain Belt of Western South Carolina, was a site of early prospecting

(California) do little to describe the professional person or the conditions under which he is forced to live; still, it is possible to rise above adversities and to profit greatly from a single discovery. But remember those funds, as large or as small as they may be must support the hunter until he finds some other source of revenue. A better approach is to engage in such activities only through alloting time from another occupation. The modern mineral explorer, sometimes called the "weekend prospector," is a new type of searcher who hopes to discover a workable deposit; but, for the most part, he recognizes the routine as a small secondary income, a happy pastime, and a source of exercise and recreation.

Basic prospection equipment

Normally, bedrock or clay bottom—not a hardpan or false bottom—is the datum plane for the prospector, since upper gravels are usually lean in gold, i. e., unless the metals are not being freshly transported across packed alluvium.

Ordinarily the bottom layer of the stream gravel becomes enriched with heavy metallic pieces (coarse gold) while the fines may be higher up. Natural selection in all cases occurs because native gold typically has a specific gravity of from fifteen to nineteen (pure gold, 19.3), while ordinary worthless rock (gangue) averages 2.6, meaning the yellow metal may be six times or more as heavy. Other substances like platinum (pure at 21.5) and diamonds, 3.5, behave similarly.

To inspect a suspect sediment, a panful of the raw sample from the bottom of the alluvium is kneaded under water until fully wetted. The pan is then horizantally rotated in small back-and-forth motions, from time to time removing the topmost gravels, and always observing the rock characteristics. When down to the remaining third, tilt the pan either forward or backward and continue in the same manner as before until a small residue is all that remains near the bottom. With the pan forward, lean it to one side and, by shaking and jiggling, gather the slow-moving gold pieces to the back of the gangue material. While everyone has his own method, the remaining tailings can be easily pulled away by wetting and sloshing the tilted residue, freeing the metal pieces. Many people at this point wash the concentrates into a second pan, continue working, and, at some later time, complete the final recovery, remembering that by counting the pans (about ten pounds for each sixteen-inch pan) and later weighing the metal from laundering gravels at several places he can, with little effort, approximate the average surface gold value per ton of ore.

Besides the value of the deposit as it appears near the surface, an experienced prospector, in observing the shape of the rocks, knows that the smooth and rounded ones were derived from an ancient stream that has moved the material a considerable distance. Angular shapes, of course, mean that the gangue is fresh, with the sharp edges thereof indicating the proximity of a vein, barren or otherwise. Equally important is the type of rock, epidote, for example, being especially interesting to gold hunters, as is unusually heavy and crystalline quartz.

A close examination of the recovered gold pieces reveals several features about a particular auriferous deposit. While size and number of metal pieces are at once important, the color (tangerine—high percentage; lemon yellow—low percentage) and particle shape are not to be overlooked. Flattened and smoothed edges indicate that the metal has been moved a considerable distance, while rough, jagged, and irregularly shaped ones with little or no evidence of being water-worn mean the contributing source is not likely to be very far away.

Specific Gravity of Some Common Materials

Osmium	22.6
Platinum	21.5
Gold	19.3
Mercury	13.5
Lead	11.3
Silver	10.5
Copper	8.9
Iron	7.9
Water	1.0

Prospecting is only a loose catch-all term that describes a wealth seeker for gold, but also for a person always on the lookout for any and every type of mineral that may show some promise. Even in these days of advanced technology, the mark of the profession, the gold pan, an object used by professionals and amateurs alike, is as simple in operation as it appears to be, probably not much different from ancient devices, and is one of the most efficient and effective methods there is for mineral exploration and evaluation; for, it is well agreed that if an experienced hunter works carefully with a pan and cannot save more than a certain percentage of the values, it is certain that his recovery is the realizable value and that mechanical devices can do no more. Clearly, to avoid disappointments in subsequent operations, a realistic value from actual recoveries by manual panning, not theoretical or assay values, is the criterion for mining potentials.

Nearly all minerals of any value have a different appearance and are heavier than ordinary gravels; therefore, the panner, in washing the rocks, is able to find different metals and several varieties of gemstones as well. One theory entertained in some circles is that wherever gold is found, diamonds may also be looked for. In large measure, diamond finds in the Southeast, as in other parts of America, are incident to prospecting and placer mining. Some agree that the geologic activity that produced the original gold deposit also played a role in producing diamonds, while others contend that washing alluvium creates a condition of exposure and observation that increases the chances of finding gemstones, especially diamonds, meaning that correlation should be with gravel volume, observation, and working time, and not necessarily with specific gold-producing localities.

Natural diamonds. The largest one is over one carat

In opposition to the theory of widespread diamond distribution, opponents claim that while the igneous bodies yielding diamonds and gold are of two different types and usually of different ages, the resulting rupture and weakening of the earth's crust by lateral forces, together with internal pressure within the earth's central molten mass, caused the magmas to flow upward into fragile regions near the surface, and for such reasons necessarily appear within the same general area. It may be that the moderate amount of diamonds found in Elliot County, Kentucky, occurs for the same reasons, though this is not yet proven. Whatever the causes, diamonds found in the Appalachian goldfields vary in size from the 1855 discovery near Manchester, Va., of twenty-four carats (twelve after cutting) through one and a half carats in South Carolina, also in 1855, to a size so small that fifty are required to weigh one carat (200 mg.).

Natural diamonds have less brilliance than the cut and polished stones; additionally they are usually out of shape (not perfect octahedrons) and may vary to almost any shade of color. Rounded forms with rough exteriors are special kinds of diamonds called bort or carbonado, and are exceedingly tough, having no fracture or so-called cleavage plane. Their resistance to shattering as a result of being fused aggregates of minute crystals while retaining the diamond hardness make them the most important industrial substance for cutting and polishing.

Of the mineral's two outstanding characteristics—luster and hardness—its adamantine brilliance sets it apart from most other stones and at once makes it recognizable as something of value. Zircon crystals, a very common mineral of the Carolina Slate Belt, closely resemble diamonds in shape and reflectance, primarily being physically distinguished by their hardness. For the most part, luster and hardness are indexed by refraction and resistance to scratching; still, the association between all four is not absolute, meaning the interatomic bonding energy in positioning the atoms may be the limiting factor. Besides greatly resembling diamonds, zircon, because of its normally minute particle size, shape, and density (specific gravity 4.7), interferes in panning,

Index of Refraction and Hardness of Some Minerals

	Refraction	*Moh's Hardness*
Diamond	2.42	10
Zircon	1.95	7.5
Sapphire	1.77	9
Spinel	1.71	8
Topaz	1.63	8
Emerald	1.59	7.5
Quartz	1.55	7
Ice	1.31	1.5

especially in the Carolina Slate Belt, making separations extremely difficult and time-consuming.

Hardness and shape of the same type of diamond varies with the locality, for weathered stones from a stream are softer than ones from a hillside alluvium, which in turn are softer than the deep underground-rock type. The rounded or irregular profile of American crystals show little if any sign of abrasion, hence they may not have been misshapen by stream wear, but rather from growth conditions at the time of formation.

1	Talc		
2	Gypsum		
		Finger Nail	2.5
		Gold	2.7
3	Calcite	Copper Coin	3
4	Fluorite		
5	Apatite		
		Window glass	5.5
6	Orthoclase	Knife	6
		File	6.5
7	Quartz		
8	Topaz		
9	Corundum		
		Silicon Carbide	14
		Boron Carbide	19.7
42.4	Diamond		

Moh's Hardness Scale (Expanded)

America's only commercial diamond mine, near Murfreesboro, Pike County, Arkansas, is a peridotite pipe similar to the Kimberly Rock formation of South Africa. Many stones were produced, once yielding a crystal of forty carats of good quality, yet the high cost of production limited the commercial output and has caused it to become a tourist attraction instead.

Nuggets and fine gold from Wolf Creek (1972)

The tried and proven method of pursuing a stream deposit is to follow the pannings by gold color, amount, size, and shape until some part of the creek shows the greatest concentration of values. At this point, the banks and hillsides are examined for the alluvial ore that produced the colors that washed into the stream. Alluvium that harbors gold will not contain the metal throughout, having routes, called pay streaks, that are part of an old creek channel crossing the gravels in nearly any direction. The final stage in prospecting, if no veins are found, is the mapping and evaluation of the pay streak.

One of the most serious problems faced by hunters with a potential gold property is in obtaining a realistically reliable value for the average precious metal content of the ore body. Field tests are adequate, provided that the sampling is properly engineered. Most miners require test holes the full depth of the entire gravel deposit, to be spaced not more than two feet apart and patterned to a grid system, some having the spacings even closer to give a clear picture of the length and breath of the auriferous sediments. If the total mineral potential is sufficient, prospecting may then give way to mining.

During the explorations, rock samples, usually quartz, may be found that may be suspected of carrying precious metals. Approximately two pounds is placed in the prospector's hand crusher (mortar and pestle) and worked to a powder. The pulverized rock is then panned as any sediment; but, in this case, the gold hunter looks for the minutest of specks to determine if the rock is totally barren. If there is any at all of any type of valuable metal, holes are dug, veins uncovered, channel samples taken, vein dip and strike measured, and the total mineral supply of the rock deposit mapped. As an added note, prospectors have learned to recognize a slight greenish tint of the gold from freshly crushed ore as being indicative of a high silver content.

As exploration develops and the hunter begins to get a clearer picture of the prospect values, suitable arrangements are made with the owner so that the prospector is definitely tied up with an option or sales agreement, for, then, he is adequately protected and all negotiations must go through his hands. Private ownership prevails in the East; free public land is available only at certain Western and Alaskan locations. Still, land titles do not necessarily mean control of mineral rights, a situation that should be well studied beforehand.

At his discretion, a prospector, to obtain exclusive rights to prospect an area for a limited time only, may require a proprietor to sign a document called a Prospector's Permit. In effect, the paper prevents encroachment by other, mineral explorers and gives the would-be profiteer first choice of mining options or mineral rights if they are made available. Conversely, the landowner may have it so worked that he is absolved from all accidents and injuries incurred during the course of the explorations.

While it is possible for miners to have overlooked some areas of promise at a particular site, more than likely the region has been thoroughly examined and what is left is of low assay and meager quantity. In that regard, it is foolhardy for a gold hunter to spend his time about an area already prospected and mined.

Furthermore, finding new lodes is becoming increasingly difficult, for the easiest have long ago been worked, leaving the lean ores and, by a very small chance, an undiscovered pocket of high-grade minerals, to the fortunate few who find hidden treasures or are able to engineer a system that will show a profit from low-value deposits.

Modern professional prospectors recognize that, even by drilling and other exploratory work, the chances of discovering a good ore body are no better than one in 1,000. One researcher reported in the *Northern Miner* (a Canadian magazine dealing in mining and minerals) that in Canada the chances of finding a metal mine are about one in 10,000.

The Miners

For some time into the foreseeable future there will be no shortage of gold deposits, unmined American reserves being estimated at sixty-nine million ounces (1972). Extracting and processing the ore at a profit, however, is a different matter altogether. Professional miners know of undeveloped, workable mineral deposits presently existing at several localities, of old mines that still contain a substantial supply of gold but which ceased production for one reason or anothe, , or of tailings from certain mines that contain sufficient precious metal value to be reprocessed for additional recovery, nevertheless, in most mining ventures large volumes are required to deliver substantial gains from the small profit margin of low recovery values for each ton processed.

Mining, as a business, involves several types of highly specialized professions; still, for someone ambitious enough to attempt to develop a resource, he could do so without being a professional, or particularly knowledgeable in any of its related fields, simply by following conventional mining techniques. For an unskilled, nonprofessional to entertain thoughts of mining, whatever the reason, the ore would almost have to be of exceptionally high-grade quality, and, equally important, subject to single ownership.

While the majority of the early mining companies did not incorporate or sell stock, careful planning of long-range ventures made it imperative that low-grade processors secure enough investment capital to gear up production for years of large-scale continuing activity. Interest in low-yield ores, such as in Arkansas, attracted gold mining companies capitalized at more than eleven million dollars from 1885 to 1888 to engage in nothing more than prospecting. In working such low-assay deposits, advantageous use of manpower and money is the only way a company can survive. In that respect, the Haile Mine in South Carolina received noteworthy attention throughout the mining industry in 1907 when 36,198 tons of hardrock ore averaging $1.49 per ton were worked at a profit.

Data on Incorporated Gold Mining Companies
in the Southern States (1902)

	Number of Companies	*Capital Stock*	*Dividends Paid*	*Assessments Levied*
Alabama	3	$1,514,000	0	0
Arkansas	1	$1,577,000	0	0
Georgia	19	$19,182,000	0	$5,000
N. C.	16	$9,884,500	0	$32,000
S. C.	2	$750,000	0	$21,000
Virginia	4	$2,600,550	0	0
	45	$35,508,050	0	$58,000

Tales of olden days are very dramatic in their portrayal of the rugged life and hard times surrounding the mining era of the 1800s and early 1900s; yet, nearly everyone is inspired by the word *gold*, hopelessly forgetting that for everyone who achieves fame and fortune there are literally hundreds of forgotten men living in despair, while thousands of others spend their entire lives at backbreaking labor at what amounts to little more than just a job.

A gold-mining facility, from a business standpoint, is no different that any other operation. According to Clyde Pittman in his book *Death of a Gold Mine*, the Haile project, the largest gold producer in the Southeast in 1902, with 175 employees averaging $1.35 per day and a fully combined facility commensurate with any other mine of its day, processed 100 tons of ore per day that on the average assayed less than $5 per ton. The monthly payroll, plus supplies and equipment, was $8,225 while the output of about 480 Troy ounces per month, at $20.67 per Troy ounce, yielded $9,921.60. This would leave $1,696.60 a month profit, or about $20,000 a year. Although no dividends were reported to have been paid to stockholders, no entry was also made for executive salaries, or federal, state, and county taxes. Even in those days of low costs, the net profit would be spread quite thin, resolving the weekly cash income to about average for the turn of the country.

That same mine, processing 400 tons a day of quartz and pyrite from open pits, in 1941, the last full year of operation, recovered 15,197 fine ounces of gold (that would be $531,895 at $35 per ounce), and 6,447 fine ounces of silver were also refined. For the state that year, a total of 15,508 Troy ounces, valued at $542,780 at $35 per ounce, was extracted. Substantial volume increases can account for more impressive gross recoveries, yet actual net profits, of course, are subject to

Central Haile Mine. The country rock, light gray schist, is in the foreground

increased wage, production, and administrative costs. Future productivity at the Haile property would include many years of operation, for core drillings of the geophysical year 1937–1939 indicated that several hundred thousand tons of gold ore at $6 a ton at $35 an ounce, still existed underground.

By contrast, gold-producing quartz stringers of the largest gold mine in America, the Homestake Mine in the Black Hills near Lead, South Dakota, with an average assay of 0.28 Troy ounces per ton ($9.80 per ton at $35 per ounce), produced 1.47 million tons in 1972, or, at 260 working days a year, nearly 5,700 tons a day. This type of volume, even with a very small profit margin, can reap large benefits, and with today's gold prices the accrued capital could be substantial; still, with a milling rate of close to one and a half million tons a year, no mine can last long. The unmined ore reserves at that site in 1972 were estimated at about 7.3 million tons averaging 0.299 ounces per ton in gold. However, because of the excessively large milling rate, that low-grade ore produced twenty-five million ounces of gold valued at $875 million from 1875 to 1960. Clearly, volume makes the difference, meaning that certain sites of the Appalachian goldfields may do as well.

Average Haile Mine Ore Assay

1883	0.55	Troy oz./ton
1883–1887	0.3	Troy oz./ton
1900	0.2	Troy oz./ton
1906	0.15	Troy oz./ton
1909-1914	0.1	Troy oz./ton
1936	0.14	Troy oz./ton
1936–1942	0.11	Troy oz./ton

Mineral recovery, particularly precious-metal mining, does not rely exclusively on assay values, but rather on a consistant yield with enough total deposit volume to allow a large-scale production to be geared up to develop the tonnage necessary to show a profit. Ores that may run $400 a ton or more and can be worked by a handful of people are rare, but not totally unknown. Usually the surface deposits that lead to the veins do so rather quickly, leading to complications that only a mining company can resolve.

Rising costs have been partly offset by improved technology and increased efficiency, yet large-scale open-pit operations still yield the lowest production costs per ton. Beyond that, placer workings are the cheapest, requiring fewer employees, and less machinery, time, and effort, therefore laundering and gravel separation is the only type of gold mining an individual with limited resources can even hope to do. In that respect, keen observation and common sense will keep the would-be profiteer from veering too far astray.

An approximately accurate estimate of the total ore value must be made before proceeding any further with ambitious scenes of sudden wealth, for all too often machines, assistants, or the type of deposit is condemned for failure to recover the metal, when the plain truth is that a sufficient quantity of gold was never there originally to justify the purchase of equipment or expenditure of time and effort. No device can create gold; in fact, it may lose a great deal through inefficiency, for none of them is better than a hand pan, only recovering more through massive volumes.

The slow recovery of fine gold at two to three minutes per pan is surpassed by the new panning champion at the international meet at Tropico Gold Mine and Camp near Rosamond, California, by recovering the eight gold nuggets placed in the sand in a new world's record time (1974) of sixteen and a half seconds; still, neither of the rates is realistic for raw panning. Alluvial ores from the Southern goldfields for the most part are hardpacked, seemingly cemented when wet or dry, early stage conglomerates (hardpans) together with clayey or kaolinoid stream bottoms requiring kneading before panning, sometimes consuming as much as five minutes prepanning time. A cement mixer, in wetting the cakey crust, works well; still, large separations will have to be accomplished if a sizeable degree of recovery is to be realized.

To equip and operate a placer at the least cost, and to show a profit against the outlay of labor and money, the initial cost of equipment must not be disproportionate to the size of the job. While experince is the best guide, a slight underrating of labor and machinery as opposed to the amount of tonnage expected may lighten later disappointments. At minimal cost, increasing yardage and saving of very fine particles augments the profit margin and may spur the miner to other sites, for there is the ever-present lure of finding, if not a bonanza, at least some unexpectedly rich ground.

Most early methods of working gravels required a constant water flow of sufficient volume flowing over riffles to trap metal pieces. Some were batch type, requiring shaking and jiggling, and functioning without mercury, while others were little more than contained artificial streams, as exemplified by sluice boxes and long toms (several sluice boxes together and extended for several yards). The constant problem of water-flow control prompted improvements and the manufacture of devices that could handle nearly any size tonnage with ease and efficiency. Even dry processers are available, albeit for small-scale work only, that function by shaking and blowing air to serve in a capacity similar to water separations.

One of the better-developed gravel-laundering machines, the Denver Mechanical Pan, one of the best-known placer machines that has found use in various parts of the world, can process twenty-five to thirty tons a day. Without the trommel and with a cement mixer, the production is considerably less. Smaller contrivances are available, or can be built that are easier to operate and less costly, yet they deliver less volume and possibily less efficiency. Using dredges in the Appalachian goldfields is not nearly as effective as would be imagined, for sediment packing in the streams prevents any outstanding volumes. To break up the encrustments just ahead of the dredge would require another device, and another person.

Most placers and some veins in the Southeast are not plagued by low values, but simply do not have enough total

Denver Mechanical Pan, shown without its trommel

volume to warrant commercial development. Even the Western and Alaskan sediment deposits are rapidly being worked out, resulting in alluvial mining, once the biggest gold-producing activity in America, now barely accounting for one percent of the total.

Gold and silver as by-products partly offset the cost of the primary production of base metals and should continue to be produced in like manner for sometime to come. Lode (vein) recovery, on the other hand, is mined for the precious metals; but the way the ores are formed, the low concentration of the metals, and the difficulties of removal from the earth, and the fact that they are even more laborious to process, requiring crushing and grinding, preliminary separations, usually roasting, flotation or cyanide treatment, and final recovery, may

Proportionate Gold Producing Methods and Total Recovery

	Lodes	*By-products*	*Placer*	*U.S. Total Recovery* (Troy oz.)
1962	47	36	17	1,542,511
1963	51	36	3	1,454,010
1964	54	37	9	1,456,308
1965	54	40	6	1,703,190
1966	58	37	5	1,803,420
1967	69	27	4	1,584,187
1968	63	34	3	1,478,292
1969	59	40	1	1,733,176
1970	60	38	2	1,743,322
1971	60	39	1	1,495,108
1972	58	41	1	1,449,943

cause the production costs to increase to such an extent as to make the recovery of America's sixty-nine million ounces of proven reserves, uncertain.

The cost of mining and milling quartz ore is from fifty to one hundred times as much per ton as the similar treatment of placer gravels; consequently, many small companies which do not have adequate facilities send their ore to toll milling plants which grind and process the rock, paying for the recovered metal at current prices minus the costs and processing fees. In that regard, Southerners have made direct shipments to custom millers at the Ducktown, Tenn. district and elsewhere; however, such toll facilities are becoming increasingly rare, with Californians presently having to send their complex ores to Arizona and as far away as Idaho.

When we reason futuristically of the Alaskan, Western, and Appalachian exploitations, and demands put upon the world's mineral resources together with their depletions, we can only conclude that long after the continents have been stripped of most of their useful minerals, the ocean will still hold great promise of wealth for those who learn to extract its incalculable mineral riches, being estimated to contain 5,688,000 tons of metallic gold alone.

Limited studies of oceanic precious metal have been

made at select sites, producing only sketchy reports as to the metal's distribution and form. No doubt some gold is in solution, probably existing as chloride, however negative results have been obtained several times from absorption tests with tin chloride, adsorption precipitation experiments, and trace recovery attempts by monel metal replacement.

It is much more probable that gold exists in ocean water as the element, and thus behaves as buoyant particles. In agreement with this, dredging sludges obtained from certain localities contain very much larger quantities of gold than the surrounding sea. Some studies indicate that the removal is accomplished by adsorption on the surface of clay sediments or even marine organisms, and is taken down by them to collect at the sea bottom.

Several determinations have shown that the distribution varies considerably from place to place and from surface to ocean floor. Information concerning oceanic gold deposits of the Southeastern coastal states is not available, yet the tropical zone of the Atlantic Ocean from top to bottom ranges from 0.004 to 3.4 milligrams per cubic meter of ocean water in metallic gold. Other median values are 0.008 mg/m^3 for the South Atlantic, 0.045 mg/m^3 for the Arctic Ocean, and 8.56 mg/m^3 off the coast of Newfoundland. The Pacific Coast near San Francisco is also above average. Some of the theories explaining precious metal occurrences in the sea include the welling action of waves and the river delivery of fine and flour gold.

According to many studies, heavy metals transported by rivers and streams are removed in quantity from the sea largely at the point of entry, resulting in less than one percent of the original amount remaining in the surrounding ocean. Fritz Haber, the famous German scientist, estimated ocean water at that low concentration to be about four tons of gold per cubic mile, later finding that the largest aggregate was in sediment ooze on the ocean floor. Recalculating the value from modern measurements reveals a retention of 1.8 tons of the precious metal in one cubic mile of sea water at an average figure of 0.4 mg/m³. Not counting ocean sediments, the value, at today's gold prices, for the metal only would be about seven million dollars per cubic mile.

Removal of Some Heavy Metals from the Sea

	Delivered into 1 m³ of the Sea	*Present in 1 m³ of the Sea*	*% Left in the Sea*
Copper	58,000 mg	10 mg	0.02
Zinc	116,000 mg	70 mg	0.06
Lead	6,000 mg	?	—
Arsenic	3,000 mg	20 mg	0.5
Gold	60 mg	0.3 mg	0.5

Several unsuccessful means have been tried in an effort to extract gold from the ocean, each achieving various degrees of success, yet none yielding enough profit to warrant continued operations. One of the more successful attempts took place in Australia where George Duncan is said to have designed a plant costing $40,000 to deal with 50,000 tons of sea water per a 24 hour day. He recovered one-tenth of a grain, or 6.5 mg., for each ton of ocean, or ten ounces per day total gold recovery. Grove Johnson repeated the test with a small plant at Portland, England, proving with similar results that, with proper methods, it is possible to mine precious metals from sea water profitably.

Even with impressive figures of high concentrations in certain places, the awesome task of processing a cubic mile of ocean water or a small fraction thereof, while not impossible, is still not yet feasible. Nevertheless, the sea, with its sediments and solutes, remains the largest storehouse of minerals of all types anywhere on earth. Consequently, with land resource depletions, the oceans will be mined and serve as another alternative to unending human needs.

Instead of outright mineral extractions, use may be made of the ability of some organisms to concentrate precious metals, either by absorption or adsorption. Additionally, at least one land vegetative plant, equisetum (horsetails or scouring rushes) concentrates gold in its seeds, yielding in the ash as much as 610 grams of metal per ton in gold-rich areas. This type of absorption and concentration may account for the reported 0.5 grams per ton of precious metal that have occasionally been reported in coal ash. Indeed, some pilot plant studies were conducted to improve metal recovery from carbonaceous gold ores using oxidation procedures based upon the usage of chlorine and sodium hypochlorite. Understandably, a precious-metal-concentrating organism that can function as a viable substitute for direct gold mining may never be found; however, as in all endeavors, every conceivable method must be considered before deciding on any one of them.

Inasmuch as Southern oceanic gold is well dispersed and may average 0.3 mg/m^3, it is at least 1,000 times as concentrated as in the earth's crust and certainly remains relatively unstudied. Equally untested is aquatic or terrestrial Appalachian plant life's ability to absorb and concentrate available precious metals. Once concentrated in lodes, however, the veins themselves, at 0.3 ounces per ton would be about 0.02 percent in metal.

Estimated Percentage of Some Common and Rare Metals in the Earth's Crust

Metal	*Percent*
Copper	0.010
Lead	0.002
Silver	0.0000004
Gold	0.0000001
Platinum	0.000000008

The Proprietors

Property titleholders are the ones most likely to profit from a mining venture, for sales, leases, options, and royalties mean revenue from a source for which there has been little proportionate capital investment. Certainly there is a wide range of variance, yet many of the present-day negotiations are conducted at rates much lower than in early mining days. North Carolina gold mines in the early 1800s were rented at various rates, the owners receiving from one-quarter to one-third of the produced value, according to the reputation of the mine. The Parker prospect, one of the first in the state, and yielding high-grade ore characteristically assaying better than 23K pure, rented for as much as half the recovery.

Col. Benjamin Hale (Haile), discoverer and owner of the Haile Mine property, was already a large landowner and wealthy planter. After his own endeavors, his property was subdivided into lots of fifty square feet and leased to various planters to work the plots, continuing for twenty years with such an arrangement. Colonel Hale probably received much less dividends from the deposit than if he had worked the ore himself; still, if he had spent most of his time and efforts at the mine he would more than likely not have become one of the most infleuntial citizens of upper South Carolina, becoming a state legislator and a trustee of South Carolina College.

In that same regard, John C. Calhoun, before becoming vice-president of the U.S., was also involved in mineral explorations, discovering and owning a substantial gold deposit about two and a half miles south of Dahlonega, Ga. Located on lots 164 and 165 of the eleventh district, adjoining the Briar Patch placer on the Chestatee River, the 600-acre tract contained one of the first sizeable mines to operate in Lumpkin County, yet it didn't make him rich, and for good or bad, things may well have worked out quite differently had he remained a miner.

Proprietors, with patience and time, may produce their own profits, with the engrossing interest dependent upon their preferences and what goals they set for themselves. The owner of the Dorn property in McCormic County, S.C., had done considerable exploring before discovering a rich ore body in 1852. Within sixteen months he had recovered $300,000 worth of gold from a surface pit 300 feet long by fifteen feet wide and twelve feet deep, at an expense that was said to be only $1,200. Company leasings followed, but after intermitent work, finally ceased after a recovery of at least $600,000.

In the Coker Creek district of Tennessee, as in many other areas of the Southeast, some of the mining was done by men who did not own the land but leased the privilege, paying a royalty of one-sixth of the placer gold recovered, but most of it was done by those owning the land who worked for gold during wet seasons, when the water was plentiful, and farmed the rest of the year.

Land ownership still means that the proprietor owns the ground as far down into the earth as any miner wishes to go; yet, before negotiating a contract two conditions should be investigated. Some individuals or organizations control the mineral rights to another person's land, while prospectors may have exclusive prospecting rights. These time-limited contracts can produce complications unless everyone is sure of the legal arrangements.

Inexpensive amateur or professional mineral explorers in no way compare to the cost and expertise of professional evaluation of land for potential mineral recovery. While most minerals have distinctive, readily observable characteristics,

Cotton Patch Mine, a well-known tourist attraction near New London, N.C.

some do not, and in many cases mines abandoned because of ore depletions of one type of mineral may contain one of another type of equal or even greater value. In that regard, Virginia's Lightfoot Mine (as with several other Southeastern prospects) is better known for its copper recovery even though it began as a gold mine and yielded a comparatively substantial volume of precious metals. Among other minerals, lead, zinc, and tungsten have at times augmented Southern gold exploitations. Moreover, pristine auxilliary ores such as bismuth and topaz may make future mining at the Brewer diggings much more attractive.

Other mines may have a sufficient volume of sericite, clays, gravels, or other minerals that may surpass the value of the precious metal that originally was responsible for the exploration. Additionally, after mine tailings have been reprocessed to remove most of the remaining metal residue, the ground rocks may be used in such areas as road building, as was the case of the mine dumps at Gold Hill, N.C. Unfortunately, such activities tend to lessen the historic value of any mining district.

40 acre rock, a large naturally exposed granitic intrusion near Pageland, S.C.

Concluding Notes

Southern gold mining may never again have the influence it once had, still it should not be thought of as a forgotten industry of the Old South. There is no evidence that present workings have generally reached the lower limit of the gold-bearing lodes, and statements by mining men in describing what is left of abandoned mines as "only roots of once mighty veins" are unfounded, for there may well be deep-lying unexplored remnants and many more vein deposits near the surface.

While some small enterprises continue to prosper, by and large prospects and hopes for extraordinary metal finds in the Appalachian gold-fields are confined to individual placer and stream workings. Certainly, known deposits of low assay exist that may have substantial volume for future productivity. Moreover, other ores may be awaiting discovery, perhaps possessing untold wealth to be found by someone courageous and daring enough to search, find, and attempt to recover precious metals from them.

Eternally optimistic would-be miners whose imaginations have been stimulated by loosely written articles should not be misled and lured into accepting as fact the notion that the ground is full of veins and placers where good wages, possibly a fortune, can be made by men of no experience who are willing but to go out and prospect. Wishing will not make it so, for obtaining gold is as much in the finding as it is in the mining. The pattern of a precious-metal-producing district anyplace,

Rolling hills of the Carolina Slate Belt, about one mile from McFarlan, N.C., and near the state line

140-Foot-Deep Open Pit at the Brewer Mine, Carolina Slate Belt

Rolling hills of the Peidmont near Jonesville, S.C., Kings Mountain Gold Belt

and especially for the pocket gold resources of the Southeast, does not mean a continuous series of ore bodies, but a spotty, hit-and-miss affair, the value of which may sometimes be boosted to a preposterous proportion.

Mark Twain, who had some experience with the exaggerations of prospectors and miners, once defined a mine as "a hole in the ground with a liar standing at the top." In that respect, experienced mineral explorers have learned not to believe anything or anyone until some proof has been shown to substantiate the claim. Novices may be well advised to exercise due caution in any mineral development scheme, especially if they are going by nothing more than what someone tells them.

That the South still holds some areas of promise that awaits discovery is undoubtedly true. The inexperienced hunter, ignorant of prospecting and mining, and hoping to find and exploit these areas profitably is certainly overoptimistic, particularly if there are little or no financial resources for commercial equipment and if he is unfamiliar with the best methods to attack inevitable uncertainties. The primitive "luck and chance" method of alluvial gold recovery, while inexpensive, yet profitable, is far removed from the scientifically planned placer and hardrock milling operations. Remember, too, that efforts to adapt improper equipment, absence of engineering experience and advice, and production on a scale too small to be economic has resulted in dissatisfaction, and, in some cases, financial ruin.

The promise of gold, from a rich mine, is one way an individual can almost immediately become wealthy, or poor; however, remember that money exchanges in metal and stone are investments that can include a wide range of minerals and of phases within each type. Gold alone involves businesses from machinery and equipment manufacturing, through smelting and refining, to bullion sales, any portion of which requires considerable funds. Moreover, any type of mineral recovery operation, or segment thereof, which is financially attractive will be pursued by some organization, meaning that many other nonglamorous minerals, because of their use and abundance, routinely surpass gold in yearly profits. The principal distinction, of course, is that an individual is sometimes able to work with precious metals alone and with a small investment, while most other mineral operations require a large capital and reliable manpower. It has been

Sand and Gravel Production in S.C., in 1972
Alone Exceeded Gold Recovery for All Its Years

	Thousands of tons	*Thousands of dollars*
1968	5,662	$ 8,074
1969	5,692	8,229
1970	5,864	7,466
1971	6,438	9,119
1972	7,916	12,121

suggested, though, that to increase the much-needed annual gold production and to utilize the excessively unemployed laboring class, a new interest in gold mining could conceivably produce as much as 350 million dollars per year (at $35 per ounce) and supply jobs to thousands of employees.

Finally, if the lure of native gold becomes irresistible, outfit for prospecting or lease the privilege for small-scale surface mining, but do not invest in any stock unless the firm is reputable and the ore is well examined for quality and quantity. In the first case, investment is small, outdoor exercise is healthy, and, if there are no outstanding finds, enough gold may be found to cover expenses. In the second case, even though stocks may be sanctioned by the Securities and Exchange Commission (established in 1934 for the protection of public investors), the rule more than the exception is that there may be sad disappointments and all manner of explanations sought as to why things turned out as they did, when in reality the original mistake was made through improper knowledge and lack of forethought.

Bibliography

Books

Armstrong, E.F., and Miall, L.M. *Raw Materials from the Sea.* New York: Chemical Publishing Co., 1946.

Boericke, William. *Prospecting and Operating Small Gold Placers.* New York: John Wiley and Sons, Inc., 1936.

Carabini, Louis, ed. *Everything You Need to Know About Gold and Silver.* New Rochelle, New York: Arlington House, 1974.

Cotterill, R.S. *The Southern Indians: The Story of the Civilized Tribes Before Removal.* Norman: University of Oklahoma Press, 1954.

Coulter, Emerton. *Georgia, a Short History.* Chapel Hill: University of North Carolina Press, 1960.

Davis, Lawton. *A History of Georgia: For Use in Schools.* New York: University Publishing Co., 1892.

Denver Processing Flow Sheets. Denver Equipment Co., Denver, Colorado.

Department of Interior. *Minerals Yearbook.* Washington, D.C.: U.S. Government Printing Office, 1941–1972.

Eaton, Clement. *History of the Old South.* New York: The Macmillan Co., 1949.

Fisher, George; Pettijohn, F.S.; Reed, J.C., Jr.; and Weaver, Kenneth, eds. *Studies of Appalachian Geology: Central and Southern.* New York: Interscience Publishers, Inc., 1970.

Hammett, A.B. *History of Gold.* Kerrville, Texas: Braswell Printing Co., 1966.

Hansen, Harry, ed. *Colorado, A Guide to the Highest State.* New York: Hastings House, 1959.

Kettell, Thomas. *Southern Wealth and Northern Profits.* University: University of Alabama Press, 1965.

Leckie, George. *Georgia, A Guide to Its Town and Countryside.* Atlanta: Tupper and Love, 1954.

Lefler, Hugh, and Newsome, Albert. *North Carolina, The Hisiory of a Southern State.* Chapel Hill: University of North Carolina Press, 1963.

Massey, J.E. *America's Money: The Story of Our Coins and Currency.* New York: Thomas Y. Crowell Co., 1968.

Payne, Dr. Henry. *The Undeveloped Mineral Resources of the South.* Baltimore: Fleet-McGinley, Inc., 1928.

Pearl, Richard. *Rocks and Minerals.* New York: Barnes and Noble, Inc., 1956.

Perkerson, Medora. *White Columns in Georgia.* New York: Holt, Reinhart and Winston, 1952.

Pittman, Clyde. *Death of a Gold Mine.* Columbia, S.C.: R.L. Bryan Co., 1972.

Pryor, E.I. *Mineral Processing.* New York: Elsevier Publishing Co., 1965.

Robinson, Blackwell. *The North Carolina Guide.* Chapel Hill: University of North Carolina Press, 1955.

Stevens, O.B., and Wright, R.F. *Georgia, Historical and Industrial.* Atlanta: Franklin Printing and Publishing Co., 1901.

Stone, Howard, and Stone, Lucille. *A Pictorial History of American Mining.* New York: Crow Publications, 1970.

Wheeler, John. *Historical Sketches of North Carolina, 1584–1851.* New York: F. H. Hitchcock, 1851.

Wilkins, Thurman. *Cherokee Tragedy: The Story of the Ridge Family and the Decimation of a People.* New York: The Macmillan Co., 1970.

Yeoman, R.S. *A Guide to U.S. Coins.* Racine, Wisconsin: Western Publishing Co., 1975.

Bulletins, Pamphlets, and Notes

Ashley, George. "The Gold Fields of Coker Creek, Monroe County, Tennessee." In *The Resources of Tennessee*, Geological Survey of Tennessee, Nashville, Tennessee, July 14, 1911.

Born, Kendall. "Summary of the Mineral Resources of Tennessee." Nashville: Division of Geology, 1936.

Bowen, R.D. "How to Get Fat on Lean Ore." *California Mining Journal*, March, 1975.

"The Bullionaires." *California Mining Journal*, reprinted from Green's Commodity Market Comments, February, 1973.

Carpenter, P. Albert, III. "Gold Resources of North Carolina." Raleigh, N.C.: Department of Natural and Economic Resources, Circular 21, 1972.

Correspondence Letter, Bureau of Geology and Topography, Trenton, New Jersey, July 11, 1975.

Daniel, Thomas, Jr.; Neathery, Thornton; and Simpson, Thomas. "Rocks and Minerals of Alabama." University, Ala.: Geological Survey of Alabama, Circular 38, 1966.

Department of the Treasury. "Coinage Documents." Washington, D.C.: Office of the Director of the U.S. Mint, July 10, 1975.

Division of Geology. "Gem Stones of South Carolina." Bulletin No. 30. Columbia, S.C.

Furcron, A.S. "The Common Rocks and Minerals of Georgia." Atlanta: Department of Mines, Mining, and Geology, Circular 5, 1961.

Grant, R. W. "Mineral Collecting in Vermont." Montpelier: Vermont Geological Survey, Special Publication No. 2, 1968.

Jett, John· "Arizona Firm Accepting Lead-Zinc Complex Ores on Toll Milling Basis." *California Mining Journal*, May, 1975.

McCauley, Camilla, and Butler, J. Robert. "Gold Resources of South Carolina." Columbia, S.C. : State Development Board, Bulletin 32, 1966.

"New Open and Junior Record Time in World Gold Panning Events at Tropico Gold Mine." *California Mining Journal*, April, 1975.

O'Neill, James, and Vallely, James. "Gold Values and Samples from the Devil's Backbone Area." University, Ala.: Geological Survey of Alabama, Circular 67, 1970.

Pardee, J.T., and Park, C.F., Jr. "Gold Deposits of the Southern Piedmont." Washington, D.C.: U.S. Geological Survey, Bulletin 213, 1948.

Reed, J.C., Jr., and Reed, J.C. "Gold Veins Near Great Falls, Maryland." Washington, D.C. : U.S. Geological Survey, Bulletin 1286, 1969.

Rose, Tom. "All About Gold." *California Mining Journal*, May, 1973.

Spaine, Suzanne. "Base and Precious Metal and Related Deposits of Alabama." University, Ala.: Geological Survey of Alabama, Circular 55, 1969.

Sweet, Palmer. "Gold Mines and Prospects in Virginia." *Virginia Minerals*. Charlottesville, Va.: Division of Natural Minerals, August, 1971.

Williams, Norman. "Mineral Resources of Arkansas." Little Rock, Ark.: Arkansas Geological and Conservation Commission, Bulletin 6, 1959.

Index